Samuel Hubbard Scudder

Historical Sketch of the Generic Names Proposed for Butterflies

A contribution to systematic nomenclature

Samuel Hubbard Scudder

Historical Sketch of the Generic Names Proposed for Butterflies
A contribution to systematic nomenclature

ISBN/EAN: 9783337013165

Printed in Europe, USA, Canada, Australia, Japan

Cover: Foto ©ninafisch / pixelio.de

More available books at **www.hansebooks.com**

HISTORICAL SKETCH

OF

THE GENERIC NAMES PROPOSED FOR BUTTERFLIES

A CONTRIBUTION TO SYSTEMATIC NOMENCLATURE

BY

SAMUEL H. SCUDDER

From the Proceedings of the American Academy of Arts and Sciences, Boston,
Vol. X. (2d S., Vol. II.)

SALEM
NATURALIST'S AGENCY
1875

HISTORICAL SKETCH

OF

THE GENERIC NAMES PROPOSED FOR BUTTERFLIES

A CONTRIBUTION TO SYSTEMATIC NOMENCLATURE

BY

SAMUEL H. SCUDDER

FROM THE PROCEEDINGS OF THE AMERICAN ACADEMY OF ARTS AND SCIENCES, BOSTON, VOL. X. (2D S., VOL. II.)

SALEM
NATURALIST'S AGENCY
1875

V.

HISTORICAL SKETCH OF THE GENERIC NAMES PROPOSED FOR BUTTERFLIES:

A CONTRIBUTION TO SYSTEMATIC NOMENCLATURE.

By Samuel H. Scudder.

Presented, Nov. 11, 1874.

> Botanicus mihi hic dicitur is, qui genera naturalia observare intelligit. Botanici (nec minus Zoölogici) autem nomine indignum judico Curiosum, qui de generibus sollicitus non est. — Linné, *Philos. botan.*
> Nomina si pereunt, perit et cognitio rerum. — Fabricius, *Philos. entom.*

Three years ago, in preparing my Systematic Revision of North American Butterflies, I first became fully aware of the extraordinary diversity of use of certain generic names in this group of insects; and I endeavored, by an historical study of the subject, to satisfy my own mind of the proper manner in which they ought to be used. The results of this study were published in the paper alluded to; but in only a few cases, and then in the briefest manner, was the process stated by which a conclusion was reached. A month or so before the issue of that paper, the late Mr. G. R. Crotch published in the Cistula Entomologica the results of an exactly similar study, based upon the same principles, but confined to an examination of those genera of butterflies which had been proposed previous to the publication of Hübner's Verzeichniss bekannter Schmetlinge. The process was in this case given, but, as it seems to me, by an unsatisfactory method, and one in which the individual opinion of the author often affected the result without the reader's cognizance.

My own paper was prepared under very unfavorable circumstances; and I therefore determined to revise its conclusions *de novo*, and to extend the study to the entire group of butterflies, as the only way in which accuracy and precision could be attained. The result is given in the present paper. The historical method is chosen as the most satisfactory one, the use of each generic name being traced from its first proposal down to the year 1874. The entire body of entomological literature has been searched with great care, and it is believed that very little of importance has escaped examination: at the same time, so much only is published as seems necessary to an elucidation of the subject.

The plan pursued with each generic name in this essay is to give, in the first place, its date, author, and place of publication, and a list of the species first included in it. For the sake of uniformity and readier comparison, these specific names (as well as all subsequent specific names) are reduced to the nomenclature of the last general catalogue of butterflies,* without which it would have been nearly impossible to have undertaken this study with the hope of any satisfactory result. Where the specific name used by the author quoted differs from the one employed for the species by Kirby, it is placed in a parenthesis, after Kirby's name; thus, in quoting the species placed by Hübner under the generic name Brangas, we have: Caranus (Pelops, Caranus), Didymaon (Dydimaon). Syncellus, Bitias. The names, as given by Hübner, stand: Pelops, Caranus, Dydimaon, Syncellus, Bitias. As reduced to Kirby's nomenclature, they are: Caranus, Didymaon, Syncellus, Bitias, Hübner's first two species being considered as one. If one or more species are indicated as types by any author, these are stated.

In a similar way, the treatment of the group by the next author is given, whose action in any manner affects its boundaries; but, in this and in subsequent cases, complete lists of the included species are not quoted, but only such a statement given as is necessary for the case in point. Other references follow, as far as they are needed, in chronological order, the dates placed at the extreme left. The action of the different authors quoted is then criticised, conclusions drawn, and attention directed to the species, which, whether from the original author's action, or by the treatment of the name by subsequent writers, should be considered as typical. For readier consultation, they are also distinguished from others given in the primary list by the use of bold-faced type in those cases where the generic name stands, or of italics where it falls; often this is the only indication of my own judgment.

Generic names which cannot be used for butterflies are followed by an asterisk.

Where the name of an author occurs in brackets, it indicates that

---

* W. F. Kirby, A Synonymic Catalogue of Diurnal Lepidoptera, London, 1871, pp. 690.

the fact of authorship is not distinctly stated, but is gathered from the context, or from subsequent works.

Names of genera which contain no butterflies are introduced wherever their members were originally considered as butterflies by the founder.

With regard to the principles upon which this work has been undertaken, I adopt, in general, those regarding genera enunciated by Agassiz in the preface to his Nomenclator Zoölogicus, and more recently by Thorell, in his work on European Spiders, with such exceptions or modifications as are indicated in my canons of systematic nomenclature.* There are, however, a few points which need special mention.

Only those names are introduced which are connected with the binomial nomenclature founded by Linné: for this reason, the trinomials of Hübner and the terms applied by Linné himself to the groups into which he divided Papilio, as well as the similar terms used by other earlier writers, such as some of those of Fabricius, Herbst, etc., have been totally disregarded. All, or nearly all, the trinomials of Hübner (used principally in the first volume of his Sammlung Exotischer Schmetterlinge, and in his Systematisch-Alphabetisches Verzeichniss) are actually used by him in some work or other (as in the Tentamen or Franck's Catalogue) with a binomial application; and in those cases they are here introduced, but only dating from the time at which and for the species for which they were employed binomially. With regard to the so-called subgeneric appellations of Linné and others, such as Plebeius, Nymphalis, etc., there are but two views which, it seems to me, can consistently be taken of them: one, that these authors always used them in a trinomial or quadrinomial nomenclature, exactly similar to that of Hübner, such as Papilio Danaus candidus rapæ, — in which case they ought not to be adopted, or else candidus should demand the same right as Danaus; the other, that they should be retained as names of groups exactly as they were first used, at the head of divisions, in a plural form, — Plebeii, Nymphales, etc. Plural nouns as titles of groups, and singular nouns with a generic signification, cannot be derived from one and the same source. " Nomina generica cum classium et ordinum naturalium nomenclaturis communia, omittenda sunt." Now the early authors, in referring to the true " genera" of Linné, always used them, as Linné did, in a singular form; but when referring to the groups into which Papilio was divided, as groups, they always used them, as Linné did, in a

---

* Amer. Journ. Sc. Arts [3], iii. 348.

plural form. The heading of the butterflies was Papilio, not Papiliones; of the swallow-tails, Equites, not Eques.

That, if used at all, they should be retained in other than a generic sense, is abundantly shown by tracing the mode in which these groups of Linné, subordinate to the genus Papilio, became the divisions subsequently termed families, and more comprehensive than the genera of modern times. Even in the last century the term "families" was applied to them; for when Cramer, in 1779, in the introduction to the first volume of his great iconographic work, alluded to the classification of Linné, it was introduced in these terms: "Je donnerai ici les divisions de M. Linné, Papillons — cinq familles." Fabricius, when he first attempted in 1807 to subdivide the butterflies into numerous genera, retained the terms Papilio and Hesperia formerly used by him, greatly restricting them of course; but did not employ, in any form whatsoever, the group-names previously in use, whether those given by Linné or those established by himself, — with a single exception, where he divides Papilio into Trojaner and Achiver, just as the Equites (to which he restricts Papilio) had before been divided into Trojani and Achivi.

But it is to French writers that we must look for the greatest light upon this subject. In Cuvier's Tableau Élémentaire (1798) we find these groups of Linné, somewhat remodelled and placed under the two genera then in use, Papilio and Hesperia: the groups, as here modified, represent in the main the families of modern times. It was during the activity of Latreille that the old genera began to be more and more restricted and new genera to multiply, until, before his death and through his writings, the interrelationship of genera and families among butterflies was entirely reversed; "families" having formerly been considered divisions of "genera," while "genera" were now looked upon as divisions of "families." In the first edition of Cuvier's Règne Animal (1817), Latreille placed all the butterflies under one "genus," Papilio, subdivided into groups termed "subgenera," which, though differing greatly from the divisions of Linné, must really be considered modifications of them, brought gradually about by the progress of science; a few, too, of Linné's names are retained. In 1825, in his Familles Naturelles, the butterflies are divided into many "genera," corresponding very closely to his previous subgeneric divisions, and ranged under one "family," Diurna, exactly corresponding to Linné's Papilio. In this connection, a study of the numerous changes in classification introduced by Latreille in his different works is very instructive. I have entered into these particulars, because Messrs. Kirby and Crotch have recently endeavored to carry back some of the Latreillean genera to Linné's time, and even to insist, for the first time, upon the necessity of employing Plebeius and similar words in a generic sense and of accrediting them to Linné. It may be added that some of these subordinate names of Linné are used in what I deem to be their true signification, as names of groups, in my Systematic Revision.

Other subsidiary principles, which are employed in this essay, should be stated. A generic name founded upon that of any species intended to be included therein, or of any synonyme of such species, must fall; and if any name falls, from this or from any other cause, it should be dropped altogether in zoölogy. I have here adopted the views of biologists who allow the repetition of names in its two departments of zoölogy and botany, but no further. And no attempt has been made to discover whether the older name (under which another may fall) is in actual use or not, since in the ever-changing sentiment among naturalists, of the generic limitation of groups, this is practically impossible, and would lead to the instability of nomenclature. The author, department, and date of publication of the older name before which any generic appellation falls, has been given, whenever possible, in order that any person may, if he choose, follow out any reference for himself, here as elsewhere. If a species is designated as type of a genus whose name cannot stand, it retains that significance when a new generic name is proposed to supplant it.

By thus calling the attention of naturalists to *historical facts* (which they may interpret in any way they judge best), I hope to have done something toward introducing some degree of fixity, logic, and precision in the generic nomenclature of the group under consideration. More perhaps than any other class of animals, unless we except Mollusca, butterflies have suffered from the writings of uneducated naturalists; and it is impossible, such has become the multiplicity of names, to reduce to order the chaotic mass of facts, excepting through their patient collation and chronological exposition. If other facts are discovered by which the result is affected, they can at once be brought into proper collocation; if a wrong interpretation is given, it is the more readily seen and pointed out. The method is clear and precise, although tedious and painful in the extreme; and such is the interrelation of usage among certain names, and the heterogeneous nature of others, as often to render the study very perplexing. The result reached in some cases will surprise many entomologists, as it has myself, and in not a few instances I would gladly see a logical way out of the necessity of change among names which have had long usage; but the law of priority is, and would best be, inexorable, and the action of those who decry it would relegate our nomenclature to an increasingly chaotic condition. I therefore hold to it as of the utmost importance in nomenclature, as the very foundation of its stability. The changes now required by its strict application are solely due to its neglect in the past. No thought of objection would arise, if it were not so. Entomologists more than others have neglected this law, have frequently acted in defiance of it, and upon them its application falls, as we should expect, most severely. A strict surveillance of systematic work hereafter will render the future, it may be hoped, less fruitful in blunders than the past.

As the work is based upon a chronological order of facts, some remarks are necessary upon two points: the dates of Hübner's different works, and that of Doubleday and Westwood's Genera. The date of Hübner's Sammlung Exotischer Schmetterlinge has generally been given as 1806–37, the years during which it is supposed to have been issued. But a careful study of the internal and external evidence shows that the dates may be much more closely approximated in all cases. The first volume contains only and all those plates to which a trinomial nomenclature is appended, and with which, as such, we have here nothing to do. The third volume, or continuation of Hübner's work, must be attributed to Geyer, and dated after Hübner's death in 1826. Hübner's Index of 244 plates (including about one hundred and seventy-five species of butterflies), in which he applies a binomial nomenclature to all the species of his first volume, is dated December, 1821, and must have been published shortly after the commencement of his second volume; for he includes in the Index twenty-one species of this volume. Supposing the plates recorded in the Index, and therefore published from 1806 to 1821 inclusive, to have been issued at regular intervals, the first volume must have been completed at about the close of 1819. We may therefore, in default of more precise data, fix upon 1806–19 as the date of the first volume, 1820–21 as that of the plates of the second recorded in the Index, and 1822–26 of those not so recorded.

This work, however, is not the only one of Hübner's which requires close examination. The Verzeichniss is dated 1816, and has always been referred to under that date. But internal evidence positively disproves this, and on that account Ochsenheimer's and Dalman's works of 1816 ante-date it. The title-page and preface to Hübner's work, the latter bearing the date 21 Sept., 1816, were printed, as the paging and signature-mark show, at the same time as the first ten pages of the catalogue itself; that is, they form a part of the first signature. But the preface to the first century of the Zuträge, which bears date 22 Dec., 1818, directly refers to a work of this nature as an unpublished desideratum. Further than this, not only are all the butterflies of the first century of the Zuträge referred to by number in the Verzeichniss,* but a species figured in the second century (Lycus Niphon (Nos. 203–4) is referred to both by name and number in the Verzeichniss, page 74. Now the preface to the second century bears the date 23 Dec., 1822. If we consider this the date when the plates of that part were completed, as is probable, then we must make the same supposition of the first century, viz., — the very end of 1818; and hence page 74 of the Verzeichniss, or, in other words, its fifth signature, and all following it, could not have been printed before two years after the Verzeichniss is dated. On page 312 of the Verzeichniss are references by number to the Zuträge, Nos. 395–6 and 429–30 the former on the last page of the second century, and the latter on the twelfth page of the third century, which dates from 27 Aug., 1825. Supposing, as before, that the preface of each part was not printed until the engraving of its plates was completed (which makes the least discrepancy), we cannot put an earlier date to page 312, or the twentieth signature, than 1823. It is questionable whether we can be so lenient as this; for it is stated by Geyer in Thon's Archiv (I. 29–30) that Hübner prepared Franck's Catalogue late in 1825. In this sale catalogue (p. 100) a list of the works of Hübner and other entomologists is given with prices annexed; and among them appear eighteen signatures (Bogen) of the Verzeichniss, probably all published at that time. We may therefore fairly conclude that, while this work was commenced in 1816, it was issued in signatures; that by the end of 1818 only the first five signatures were printed, and by the end of 1822 only the first twenty. More probably, however, only the first eighteen signatures were printed before the autumn of 1825. The work was completed by Hübner and wholly published by 1827, judging from Geyer's list of Hübner's works given in Thon's Archiv † (l. c.). Doubtless a

---

* Excepting only Nos. 193–4, which are not referred to at all; and a few of the later ones, which are referred to by name only, — viz., Nos. 163–4 on page 9 of the Verzeichniss, 187–8 on page 11, 188–90 on page 80, and 107–8 on page 47.

† The price of the work is given there as 44 kreutzers only, while that of the Syst.-alph. Verzeichniss, not one-fifth its size, as 54 kreutzers. This may probably be accounted for by the greater rarity of the latter, rather than by an incomplete condition of the former.

nearer approach could be made toward the dates of the different parts of the book by a comparison of the moths with those of the Zuträge. These facts are given to show that the whole work could not have been published in 1816. Still, for mere convenience and uniformity, I have used 1816 as the date; for the only case where the dates conflict with those of another writer in the use of the same generic name is that of Eurybia, which should unquestionably be referred to Illiger.

The preceding statement also shows that the dates of the different parts of the Zuträge are probably correct.

The Tentamen* is undated. It is twice referred to by Hübner himself: once in the preface to his Verzeichniss, written in 1816; and again, in 1818, in the preface to the first century of his Zuträge. In the latter case it is not specified by name, but the substance of it is reprinted, and there is no other work of Hübner's to which his words can refer; it is stated to have been published in 1806. It is also referred to by Ochsenheimer in 1816, in the preface to the fourth volume of his Schmetterlinge Europas, as having been unknown to him at the time of the publication of the first volume of the same work, in 1807; it is also included by Geyer in his list of Hübner's works, and by Hagen in his Bibliotheca Entomologica.

I am greatly indebted to Dr. Hagen, of Cambridge, and to Herr Gerichtsrath Keferstein, of Erfurt, for their kind assistance in my endeavor to discover the dates of Hübner's works. It would be a worthy task, if one of the Berlin entomologists would examine the works of Hübner in the Königliche Bibliothek, where, I am told by Dr. Hagen, they are preserved in their original wrappers.

There is still another work, the dates of the different parts of which, as given here, require explanation. Doubleday and Westwood's Genera of Diurnal Lepidoptera was published in parts, and Mr. B. P. Mann has shown me a nearly complete set of the work in the original wrappers; although it is the reissue and not the original edition, a careful comparison of its divisions with the dates printed at the bottom of many of the signatures, convinces me that the reissue was purely a reissue, and that the plates accompanying each part of the text are the same as in the original issue. The dates given below are based upon this supposition.

The dates of different parts of such of Boisduval's works as appeared by livraisons are drawn from the official literary bulletin published in Paris at that time, and can be relied upon for accuracy.

---

* Republished by me in fac-simile. Cambridge, 1873.

References to Leach in Brewster's Encyclopædia are to the paging in the American edition; all the references to Hewitson's Exotic Butterflies are at second-hand.

In conclusion, I would return my thanks to many entomologists who have answered special inquiries concerning works and insects to which I had no ready access; and especially to Mr. W. F. Kirby, of Dublin, and the late Mr. G. R. Crotch, of Cambridge, with whom I have constantly consulted, and whose aid has been of the greatest importance.

1. ABÆIS.

1816. Hübn., Verz. 97: Nicippe, **Cebrene** (Arethusa), Brigitta.
1870. Butl., Cist. Ent. i. 35: designated Nicippe as type.
1872. Scudd., Syst. Rev. 39: does the same.

Nevertheless Nicippe cannot be taken as the type, for that species must be reserved for Xanthidia (1829). The other species referred to it belonging to the genus Eurema (1816), Cebrene may be taken as the type.

2. ABANTIS.

1855. Hopff. Verh. Akad. Wissensch. Berl. 643: **tettensis**. Sole species, and therefore type.

3. ABISARA.

1860. Feld., Wien. Ent. Monatschr. iv. 397: **Echerius** (Kausambi), Savitri, Damajanti.
1867. Bates, Journ. Linn. Soc. Lond., Zoöl. ix. 413: extends the genus, but includes in it only the former two of the original species.

Echerius may be considered the type.

4. ABROTA.

1858. Moore, Cat. Lep. E. Ind. Co. i. 176: **Mirus** (Ganga). Sole species, and therefore type.

5. ACAPTERA.

1820. Billb., Enum. Ins. 76: **crisia**. Sole species and designated type.

6. ACCA.

1816. Hübn., Verz. 44: Melicerta (Blandina), Agatha, Columella (Columena), aceris (Matuta, aceris), Sappho (Lucilla), **Venilia**, Heliodora, Lucothoe, Ophione, Valentina, Sulpitia, Hera.

1865. Herr.-Schaeff., Prodr. i. 66: confines the genus to two species, Procris and Urdaneta, not mentioned at all by Hübner, nor very closely related to the original types, but placed by Kirby in the genus Limenitis. They have therefore nothing to do with Acca.

Felder in his Neues Lepidopteron divides the genus Neptis into seven sections, the fifth containing the species Venilia. It is to this group that I would restrict Hübner's generic name Acca, with Venilia for type. See Procris.

7. ACENTROCNEME.

—— Feld. MS., in a copy of Feld., Lep. Fragm. 46: **Kollari**. Sole species, and therefore type.

Proposed by the author, in a copy of his work sent to Frauenfeld, as a substitute for Ægiale (q. v.) preoccupied. I do not find it published anywhere.

8. ACHALARUS.

1872. Scudd., Syst. Rev. 50: **Lycidas**. Sole species and designated type.

9. ACHILLIDES.

1816. Hübn., Verz. 85: Bianor, **Paris**, Helenus, Severus, Deiphobus (Deiphobus, Alcandor), Agenor* (Achates, Alcanor).

Paris may be considered as the type.

10. ACHLYODES.

1816. Hübn., Verz. 107: Busiris (Busiris, Bucolus), Thraso, **Fredericus** (Fridericus).

1852. Westw., Gen. Diurn. Lep. 524: employs it for several species, including, of Hübner's, all excepting Busiris.

1869. Butl., Cat. Fabr. Lep. 285: employs it for several species, including, of Hübner's, only Busiris.

1870. Ib., Ent. Monthl. Mag. vii. 98: designates Busiris as type, but wrongly, on account of Westwood's action.

1872. Scudd., Syst. Rev. 50 [Achylodes]: does the same, with similar error. See also Helias.

The three species placed by Hübner in this group belong to as many genera; and therefore, as Thraso was taken as type of Eantis (q. v.) in 1836, Fredericus must be the type of this.

11. ACHNA.*

1820. Billb., Enum. Ins. 80: proposed for Helias, for no reason whatever. Helias (q. v.) falls for lack of members.

---

* Placed also by Hübner, in the same work, in Iliades.

12. ACIDALIA.

1816. Hübn., Verz. 31 : Cybele, Aglaia (Aglaja), Adippe, Niobe, **Niphe.**
1850. Steph., Cat. Br. Lep. 13, 258 : uses the name for the three middle species, Aglaja, Adippe, Niobe.
1858. Kirb., List Br. Rhop. : employs it for the same and others.

These, however, as well as Cybele, all belong to the earlier Argynnis. Niphe may therefore be taken as the type.

13. ACOLASTUS.

1872. Scudd., Syst. Rev. 50 : **Savignyi.** Sole species and designated type. See Polygonus.

14. ACONTHEA.*

1829. Horsf., Descr. Cat. Lep. E. Ind. Co. (expl. plates) : Adonia (Lubentina), Alankara, Aconthea (primaria).
1829-30. Ib., Zoöl. Journ. xvii. 65 : Cocytina, Coresia (Apaturina), Nero (Thyria).

Being founded upon the name of one of the species included in the group,* the name falls, and cannot properly be used ; moreover, the name Acontia (Hübn., Lep. 1816) is, perhaps, too closely allied. See Adolias.

15. ACONTIA.*

1847-48. Westw., Cab. Or. Ent. 76, pl. 37 : *Siva* (Doubledayi). Sole species, and therefore type.

The species has, however, been made the type of the genus Neurosigma (q. v.), and the generic name is preoccupied in Lepidoptera (Hübn. 1816).

16. ACRÆA.

1807. Fabr., Ill. Mag. vi. 284 : **Horta,** Terpsichore, Bellona (Brassolis).

The first two species are Acræans, as understood in recent times ; the last, however, is a Pierid.

1816. Hübn., Verz. 92 : places five species under this generic name, none of which have any thing whatever to do with the Fabrician group ; most of them are Pierids. His genus Telchinia corresponds in general to the Fabrician Acræa. Later authors have retained the Fabrician name for this group.
1872. Crotch, Cist. Ent. i. 66 : specifies horta as type.

---

* The first citation is undoubtedly the earlier.

17. ACROPHTHALMIA.

1861. Feld., Wien. Ent. Monatschr. v. 305 [Acrophtalmia]: **Artemis.** Sole species, and therefore type.
1867. Ib., Reise Novara, 486: corrects the name to Acrophthalmia, and it is so used by Kirby.

18. ACTINOTE.

1816. Hübn., Verz. 27: Thalia, Gea (Epæa), **Euryta** (Eurita), Amosis (Amesis).
1848. Doubl., Gen. Diurn. Lep. 142: retains it for Thalia and seven others, placed in two sections.
1869. Butl., Cat. Fabr. Lep. 128: employs it for Thalia only.

    Nevertheless Thalia cannot be designated as the type (see Calornis); nor can Amosis, because it was placed in Alesa in 1847. Euryta may be taken as the type.

19. ACULHUA.

1871. Kirb., Syn. Cat. Lep. 301: **Cinaron.** Sole species, and therefore type.

    The name is proposed in place of Dryas Feld., nec Boisd., nec Hübn.

20. ADELPHA.

1816. Hübn., Verz. 42: **Mesentina,** Basilea (basilis), Iphicla, Plesaure, Cocala, Cytherea (Elea, Cytherea), Phliasus (Phliase).
1865. Herr.-Schaeff., Prodr. i. 66: employs it for Irmina and five others, of which only Mesentina (Mesenteria) is mentioned by Hübner.
1871. Kirb., Syn. Cat. Lep. 230: employs it for all the species mentioned by Hübner and Herrich-Schaeffer, excepting the last of Hübner's, which is wrongly placed in this connection.

    Mesentina may be designated as type.

21. ADOLIAS.

1836. Boisd., Spec. gén., plates 3, 4 B.: **Aconthea** [larva only], Dirtea (Boisduvalii).
1844. Doubl., List Br. Mus. 102: places a number of species in the genus, among them Aconthea.

1850. Westw., Gen. Diurn. Lep. 289: places twenty-five species in the genus, among them Aconthea, which he specifies as type.*

1861. Feld., Neues Lep. 34: divides the genus into ten sections, the first of which he names Itanus, and places in it Aconthea and four others. See Aconthea.

22. Adopæa.

1820. Billb., Enum. Ins. 81: **Thaumas** (linea) and a MS. species.
Thaumas is therefore the type. See Pelion.

23. Ægiale.*

1860. Feld., Wien. Ent. Monatschr. iv. 110: *Kollari*. Sole species, and therefore type.

This generic term is too close to Ægialia (Latr., Col. 1807); and probably for this reason in a copy of the Lepidopterologische Fragmente in my possession the name is erased, and Acentrocneme (q. v.) substituted.

24. Æmona.

1868. Hewits., Exot. Butt. iv. 64: **Amathusia**. Sole species, and therefore type.

25. Æola.*

1820. Billb., Enum. Ins. 78: Iris, Ilia, Bolina (Lascinassa, Bolina), and a MS. species.

No matter which species is chosen as the type, the genus is preoccupied. See Apatura and Potamis.

26. Aeria.

1816. Hübn., Verz. 9: Nasica, Reckia (Reckii), **Aegle**, Eumelia (Vocula), assarica (asarica).

The first species is a moth, and Eumelia is very distinct from the others.

1844. Doubl., List Br. Mus. 149: places six species in this group, among which are Aegle and Reckia of Hübner's list.

1847. Ib., Gen. Diurn. Lep. 126: places four species in the group, of which Aegle is the only one of the original species of Hübner's.

Aegle then should be the type. See Choridis.

---

* It may seem out of place to some to consider a species as type, when reference is originally made to the larva only; but the entire force of the objection is lost, when we remember that generic distinctions are as easily traced in the larva as in the imago.

27. ÆRODES.

1820. Dalm. in Billb., Enum. Ins. 79: **Idomeneus**. Sole species, and therefore type.

    If, however, this species is strictly congeneric with Eurylochus, the genus will fall before the earlier Caligo (q. v.).

28. ÆROPETES.*

1820. Billb., Enum. Ins. 79: Licus (Licas), Tulbaghia.

    There is a Castnian with the name of the first species, and it is probably the insect meant by Billberg, although the species is Drury's, and not Fabricius's, as stated by the writer. The group as thus constituted consists of wholly incongruous material, and may be discarded. See Meneris.

29. ÆTHEIUS.

1816. Hübn., Verz. 109: Pretus, **Archytas**, Meris.

    Archytas may be selected as type, although belonging to a different family from the other two; for it alone belongs to the group in which Hübner placed this genus.

30. ÆTHILLA.

1868. Hewits., Hesp. 55: **Eleusinia.*** Sole species, and therefore type.

1870. Butl., Ent. Monthl. Mag. vii. 57: designates Eleusinia as type.

31. AGANISTHOS.*

1836. Boisd., Spec. gén., pl. 4 B.: *Odius* (Orion). Sole species, and therefore type.

    Used in same sense by subsequent authors. Probably, however, it must fall before Historis (q. v.)

32. AGAPETES.

1820. Billb., Enum. Ins. 78: **Galathea**, Lachesis.

    Galathea may be taken as type. See Melanargia, Satyrus, and Arge.

33. AGATIINA.*

1843. White, Zoöl. i. 28: *Margaretta*. Sole species, and therefore type.

    The name is, however, preoccupied in mollusks (Raf. 1831).

34. AGERONIA.

1816. Hübn., Verz. 42: Amphinome, Arethusa (Laodamia), Feronia, Chloe.

    Subsequent usage has been in accordance with this.

1861. Feld., Neues Lep. 17: divides the genus into four sections, the fourth of which, unnamed, contains only the last species mentioned by Hübner.

Chloe may therefore be considered as the type. See Peridromia.

35. AGLAIS.

1816. Dalm., Vetensk. Acad. Handl. xxxvii. 56, 64: Io, Antiopa, Polychloros, urticæ, c. album, Atalanta, cardui; urticæ specified as type.
1872. Scudd., Syst. Rev. 16: also specifies urticæ as type.

36. AGLAURA.*

1851. Boisd. in Westw., Gen. Diurn. Lep. 327: Westwood gives this as a MS. synonyme of Zeuxidia (q. v.).

It is preoccupied in Acalephs (Pér.-Les. 1809) and Worms (Sav. 1817).

37. AGRAULIS.

1833–4. Boisd.-LeC., Lép. Am. Sept. 142: **vanillæ**. Sole species, and therefore type.
1836. Boisd., Spec. gén., pl. 6 B.: Moneta.
1861. Feld., Neues Lep. 7: separates two sections, the first including vanillæ and Juno, the second Moneta.

38. AGRIADES.

1816. Hübn., Verz. 68: Endymion (Daphnis), Laius (Cajus), Panoptes, Argiolus, Ladon, Admetus, **Orbitulus**, Corydon, Dorylas (Dorylas, Golgus), Thetis (Adonis), Alexis (Agestis), Chiron (Eumedon), Icarus (Icarius).
1850. Steph., Cat. Br. Lep. 19, 261: places in it Corydon, Thetis (Adonis), Alexis, Dorylas, and Icarus (Icarius, Eros).
1858. Kirb., List Br. Rhop.: places in it Argiolus, Corydon, Alexis (Agestis), and Artaxerxes (Salmacis, Artaxerxes).

The species mentioned by Stephens and Kirby seem to belong to the earlier Rusticus, and hence have no effect. Orbitulus may be taken as the type.

39. AGRIAS.

1844. Boisd. in Doubl., List Br. Mus. 106: **Claudia**, Blomfildia (Blomfildia, bella).
1848. Boisd. MS. by Doubl. in Hewits., Proc. Zoöl. Soc. Lond. xvi. 45: Ædon.

1850. Westw., Gen. Diurn. Lep. 298 : Claudia, Ædon. He credits Boisduval with the name, but restricts the group to one of his MS. sections.

1870. Boisd., Lép. Guat. 52: claims the name, and refers Ædon to it.

1871. Kirb., Syn. Cat. 265: uses it for Claudia, Ædon, and others.

> Claudia may be considered as the type through Westwood. The name is rather close to Agria, used in Diptera (Rob.-Desv. 1830).

40. AGRODIÆTUS.

1825. Hübn., Catal. Franck, 82: Semiargus (acis), Cyllarus (Damœtas), Argiolus, Iolas, Damon, Endymion (Daphnis), Arcas (Erebus), Corydon, Orbitulus, Icarius, Dorylas, Thetis (Adonis), Icarus (Alexis), Argus, Hylas, Optilete, Argiades (Polysperchon), Bœticus, roboris (Evippus), Virgaureæ, Gordius, Thersamon, Phlæas, Ballus, Hippothoe (Chryseis), Alciphron (Hipponoe), Spini, Strephon (Sicheus), Quercus, W. album, ilicis (Lynceus), Beon, Eurytulus, Hemon (Hemon, Acmon), Atys, Marsyas, betulæ (betuli), imperialis (Venus), Helius (Eurisus).

> Damon may be taken as the type.

41. AIDES.*

1820. Billb., Enum. Ins. 81 : Epitus (Epithus), Phocus (Phocas), Proteus.

> This name is preoccupied through Aidos (Hübn., Lep. 1816).

42. AILUS.*

1820. Billb., Enum. Ins. 81: proposes, without reason, to use this name for Zelima (q. v.).

43. AJANTIS.

1816. Hübn., Verz. 13: **Sappho**, Antiochus (Antiocha), Pasithoe (Hecale).

> Sappho, which is generically distinct from the others, may be taken as the type.

44. ALÆNA.

1847. Boisd., Voy. Delag. ii. 591: **Amazoula**. Sole species, and therefore type.

45. ALAZONIA.*

1816.   Hübn., Verz. 46: Cydippe, Cyane (Penthesilea, Simbiblis).

Unless Cyane should prove generically distinct from Cydippe, as scarcely seems probable, this name must fall before Cethosia (q. v.).

46. ALCIDIS.*

1860.   Feld., Wien. Ent. Monatschr. iv. 250: *Liris*. Sole species, and therefore type.

But the name is preoccupied in Lepidoptera (Hübn. 1816). [See Appendix, p. 293.]

47. ALCYONEIS.*

1816.   Hübn., Verz. 35: Asterie, Almana (Almane).

This name falls before Junonia of the same author, both its species being generically identical with those of Junonia.

48. ALESA.

1847.   Doubl., List Br. Mus. 1: Amosis (Priolas), **Prema**.

Subsequent authors (Westwood, Bates, Kirby) having always placed Prema first on the now more extended list of species, it may be considered as type.

49. ALGIA.*

1865.   Herr.-Schaeff., Prodr. i. 77: *Satyrina*. Sole species, and therefore type.

But the species is inedited and the genus undescribed, its place only indicated as between Lachnoptera and Messaras; consequently the name must be dropped.

50. ALLOTINUS.

1865.   Boisd. in Feld., Reise Novara, 285: **Fallax**, major, subviolaceus, unicolor, albatus.

The first species being the only one credited to Boisduval, that must be considered the type.

51. ALCŒIDES.

1816.   Hübn., Verz. 73: Thyra, **Pierus**.

Pierus may be taken as the type.

52. AMARYNTHIS.

1816.   Hübn., Verz. 26: **Meneria** (Menaria). Sole species, and therefore type.

The genus has always been used in this sense.

53. AMARYSSUS.*

1816. Dalm., Vetensk. Acad. Handl. xxxvii. 60, 85: *Machaon*.
Sole species and designated type.
1820. Billb., Enum. Ins.: applies it wrongly to other swallow-tails.

But Machaon had earlier been specified as type of Princeps, and therefore this genus falls, and cannot again be employed. See Papilio.

54. AMATHUSIA.

1807. Fabr., Ill. Mag. vi. 279: **Phidippus**. Sole species, and therefore type.

It has always been used in this sense. See Mitocerus.

55. AMAURIS.

1816. Hübn., Verz. 14: **Niavius** (Niavia), Egialea, Echeria.
1866. Reak., Proc. Acad. Nat. Soc. Philad. 33: uses it in the same sense, adding another species.
1871. Kirb., Syn. Cat. 8: employs it similarly.

Niavius may be considered as the type.

56. AMBLYGONIA.*

1865. Feld., Reise Novara, 308: Eumæus (Agathon), Amarynthina.
Falls before Notheme, and is preoccupied (Herr.-Schaeff., Lep. 1855).

57. AMBLYPODIA.

1829. Horsf., Descr. Cat. Lep. E. Ind. Co. 98: I. Narada; II. Vivarna; III. **Apidanus**, Centaurus, Ædias (Helus), Eumolphus; IV. Phocides (Sugriva); V. Vulcanus, Lohita, Syama, Timoleon (Rochana), Jalindra, Longinus, Erylus, Jangala, Vidura, Etolus.
1847. Doubl., List Brit. Mus. 23: uses it for Narada, Apidanus, Centaurus (Pseudocentaurus), Eumolphus, Timoleon (Rochana), Longinus, Jangala, Vidura, and others which are mostly MS. species.
1852. Westw., Gen. Diurn. Lep. 477: employs it for all these and others, specifying Centaurus, Apidanus, Ædias (Helus), and Anthelus — the last only not previously mentioned — as the types.
1868. Herr.-Schaeff., Prodr. ii. 18: gives Narada and two others.
1870. Boisd., Lép. Guat. 14: specifies Narada as the type, but incorrectly, through Westwood's previous limitation.
1871. Kirb., Syn. Cat. 419: employs it for a large number of species, including the four types mentioned by Westwood.

Apidanus may be taken as the type.

58. AMBLYSCIRTES.

1872. Scudd., Syst. Rev. 54: **vialis**, Hegon (Samoset), Tolteca. The first species specified as type.

59. AMECERA.*

1867 (March). Butl., Ann. Mag. Nat. Hist. [3] xix. 163: *Megæra* (Megæra, Lyssa), Tigelius, Mæra, Eversmanii, Hiera, Schakra (Shakra), Menara, Baldiva.

> The author says that Dira of Hübner "cannot be adopted, as it includes several distinct forms, the type species moreover being a true Lasiommata."

1867 (June). Ib., Entom. iii. 280: Megæra.
1868. Ib., Ent. Monthl. Mag. iv. 195; Cat. Sat. 123: specifies Megæra as the type.

> The name must fall before Lasiommata (q.v.), unless some of the species first mentioned by Butler should prove to be generically distinct from Megæra; this is hardly probable. Dira of Hübner contains represeresentatives of three different genera, and can be retained for one of them.

60. AMECHANIA.*

1861. Hewits., Exot. Butt. ii. 87: *incerta*. Sole species, and therefore type.
1861. Herr.-Schaeff., Ex. Schm. pt. 39: incerta. The genus is to be credited to Hewitson, since Herrich-Schaeffer does so in his Prodromus.

> The genus, however, according to Butler, is strictly congeneric with Zethera, which has precedence by one month. Amechania must therefore drop, and cannot again be employed.

61. AMMIRALIS.*

1832. Renn., Consp. 10: *Atalanta*. Sole species, and therefore type.

> The genus falls before Vanessa. See also Pyrameis and Bassaris.

62. AMNOSIA.

1844. [Boisd. in] Doubl., List Br. Mus. 88: **decora**. Sole species, and therefore type.
1850. Westw., Gen. Diurn. Lep. 259: makes the same use of it, but accredits the generic name to Boisduval, in whose name it must therefore stand. See Leptoptera.

63. AMPHICHLORA.

1861. Feld., Neues Lep. 19 [as section of Ageronia]: **Feronia** (Feronia, Epinome), Ferentina, Fornax.

1865. Herr.-Schaeff., Prodr. i. 76: Chloe. The generic name is credited to Boisduval!
    Feronia may be taken as the type.

64. AMPHIDECTA.

1867. Butl., Ann. Mag. Nat. Hist. [3] xx. 404: **pignerator.** Sole species, and therefore type.

65. AMPHIDEMA.

1861. Feld., Neues Lep. 27: **Beckeri.** Sole species, and therefore type.

66. AMPHIRENE.*

1844. Doubl., List Br. Mus. 86: Trayja (Traja), *Epaphus.*
1848. Ib., Gen. Diurn. Lep., pl. 32: Epaphus.
1870. Boisd., Lép. Guat. 43: Epaphus (Epaphea) and others.
    This name falls before Siproeta (q. v.).

67. AMPHRISIUS.*

1832-33. Swains., Zoöl. Ill. ii. 98: *Pompeus* (Nymphalides).
    Amphrisius is one of the synonymes of this species; and the generic name being founded upon it falls, and cannot be employed. See Troides.

68. AMYCLA.*

1849. Doubl., Gen. Diurn. Lep. 223: *Taurione* and three more to which a query is attached; namely, Orphise (Orphise Triphosa), Amycla, and Cœlina.
    It is employed for Taurione and another species by Felder (Neues Lepid.); but, being founded upon the name of one of the species originally included in it, it falls, notwithstanding that Doubleday expressly says that the species Amycla may belong to Cybdelis.

69. AMYNTHIA.

1832-33. Swains., Zoöl. Ill. ii. 65: Chlorinde (Swainsonia), **Maerula** (Merula); type specified as Mærula.
1847. Doubl., Gen. Diurn. Lep. i. 70: proposes that it should be used for the American species placed by him in Gonepteryx, which includes both of the above. (See also Rhodocera.)
1870. Butl., Cist. Ent. i. 35, 45: indicates Chlorinde (Swainsonia) as type, but of course erroneously.
    This generic name must be retained, because Anteos (q. v.) is virtually preoccupied.

70. ANADEBIS.

1867. Butl., Ann. Mag. Nat. Hist. [3] xix. 50 : **Himachala.** Sole species, and therefore type, as subsequently indicated by the same writer. See also Theope.

71. ANÆA.

1816. Hübn., Verz. 48 : **Troglodyta** (Troglodita), Morvus (Laertias, Acidalia), Leonida, Rhipheus (Riphea). The last is not a butterfly.

    Troglodyta may be taken as the type.

72. ANAPHÆIS.

1816. Hübn., Verz. 93 : **Creona**, Chloris, Java (Coronea).

    Creona may be taken as the type.

73. ANARTIA.

1816. Hübn., Verz. 33 : Arsinoe, **Jatrophæ**, Amalthea (Amathea).
1849. Doubl., Gen. Diurn. Lep. 214: divides the group into two sections, in the first of which he places Jatrophæ and in the second Lytrea (Lytræa), and others, including Amalthea (Amathea); the group is equivalent, he says, to Boisduval's MS. genus Celæna (see Celœna).

    Felder adopts the same division, and we may therefore restrict the group to the first division, and consider Jatrophæ as the type.

74. ANASTRUS.

1822-26. Hübn., Ex. Schm. ii.: **Corbulo** (obscurus). Sole species, and therefore type.

    This may stand, though it is worth stating that Hübner had previously (Verz. 1816) placed this species in two other and different genera. See Celænorrhinus and Talides.

75. ANATOLE.

1816. Hübn., Verz. 24 : **Zygia**, Peuthea.

    These two species not being congeneric, and this generic term having been retained by different authors, such as Doubleday, Westwood, Bates, Kirby, for the first species, it may be considered as the type.

76. ANCHYPHLEBIA.*

1868. Butl., Ent. Monthl. Mag. iv. 195: *Archæa.* Sole species, and specified type.

    Falls before Antirrhea (q.v.); Butler's objection to Antirrhea, that it was not characterized by its author, may be urged just as strongly against many of Boisduval's genera, accepted by him.

77. ANCISTROCAMPTA.

1862. Feld., Wien. Ent. Monatschr. vi. 183: **Hiarbas** (Syllius). Sole species, and therefore type, as stated subsequently by Butler.

78. ANCYLOXYPHA.

1862. Feld., Verh. zoöl.-bot. Gesellsch. Wien, xii. 477: **Numitor,** corades. Numitor is specified as type.
1872. Scudd., Syst. Rev. 53: also specifies Numitor as type.

79. ANCYLURIS.

1816. Hübn., Verz. 23: Tedea, **Aulestes** (Pyrete), Periander (Periandra).

Kirby (Syn. Cat.) has used this term in the place of Erycina (preoccupied), placing in it the first two species, which are not congeneric, and others.

Tedea belongs to Zeonia (1832–33), Periander was taken in 1837 as the type of Diorina, and hence Aulestes must be taken as the type. See Rodinia.

80. ANDROPODUM.

1825. Hübn., Catal. Franck, 84: cratægi, **Ilaire** (Margarita), Lycimmia (Limnoria), Monuste? (Pseudomonuste), Pyrrha (Eieidias), Eucharis, Tereas, Eurota, [?] Buniæ (Endeis), cheiranthi, brassicæ, napi (napi, bryoniæ), Callidice, Anguitia, Daplidice, Belemia, Ausonia (Belia, Ausonia), Eupheno, cardamines, sinapis (lathyri), Phiale, Albula, Elathea, Delia (Daira), Nise, Croceus (Edusa), Chrysotheme, Hyale, Phicomene, Palæno, Argante (Hersilia), Philea, Eubule (Eubule, Sennæ), Cipris (Cypris), Statira (Evadne), Cleopatra, rhamni, and a MS. species.

Ilaire may be taken as the type.

81. ANELIA.

1822–26. Hübn., Exot. Schmett. ii.: **Numida** (Numidia). Sole species, and therefore type.
1827–37. Gey. in Hübn., Exot. Schmett. iii.: Thirza. See Clothilda and Synalpe.

82. ANEMECA.

1871. Kirb., Syn. Cat. 179: **Ehrenbergii.** Sole species, and therefore type. See also Morpheis.

83. Anops.*

1836. Boisd., Spec. gén., pl. 7 C.: *Thetys* (Phædrus). Sole species, and therefore type.

Since used by Doubleday and Westwood, but the name is preoccupied in Crustacea (Oken, 1815), and Reptiles (Bell, 1833). See also Curetis and Phædra.

84. Anosia.

1816. Hübn., Verz. 16: Erippus (Archippe, Erippe), Misippus (Misippe), **Gilippus** (Menippe, Vincedoxici, Eresima).

As Misippus is totally distinct from the other species of this genus as well from the group to which it belongs (having been placed here on account of its mimetic resemblance), it can in no case be considered or made the type of the genus; the other species not being strictly congeneric, and Erippus being already excluded, from its relation to Danaida, Gilippus must be taken as the type.

85. Anteos.*

1816. Hübn., Verz. 99: rhamni, Mærula, Cleopatra.

This name must be dropped, from its too close resemblance to Anteon (Jurine, Hym. 1807). See Amynthia and Colias.

86. Anteros.

1816. Hübn., Verz. 77: **formosus**, Achæus.

It has since been used (Doubleday, Westwood, Bates, Kirby) in the same sense. Formosus may be taken as the type.

87. Anthene.*

1847. Doubl., List Br. Mus. 27: Galathea, Larydas.

This term is too close to Anthenea (Gray, Echin. 1840) to be employed.

88. Anthocharis.

1836. Boisd., Spec. gén. 556: I. Belemia (Belemia, Glauce), Ausonia (Belia, Ausonia, Simplonia), Tagis, Eupheno, Damone, cardamines, **Genutia**; II. chilensis; III. subfasciata; IV. Evanthe, Eucharis, Evarne, Danæ, Eupompe, Achine (Antevippe, Achine), Antigone, Evippe, Omphale, Theogone, Etrida, Phlegetonia, Delphine, Eione, Daira, Evagore, Ephyia (Ephya), Liagore, Eulimene, Arethusa, Cebrene, Ocale.

1847. Doubl., Gen. Diurn. Lep. 56: places in Anthocharis (sens. strict.) ten species, including Belemia and Genutia.

As Euchloe (q. v.) must be used for the European species, Genutia may be considered the type of this genus. See also Midea.

89. ANTHOMASTER.

1872. Scudd., Syst. Rev. 57 : **Leonardus**, Uncas. Leonardus specified as type.

90. ANTHOPSYCHE.

1857. Wallengr., Rhop. Caffr. 10 : I. Achine, Omphale, Evenina, Procne, Phlegetonia, Gavisa; II. Eupompe, Danæ, Evarne, Eucharis, Agoye, Eris, Ione (Jone, speciosa).

We propose restricting this group to the first section, with Achine as type. For the second section, see Callosune.

91. ANTHORA.*

1844. Doubl., List Br. Mus. 99 : *Eurinome*. Sole species, and therefore type.

This generic name falls before Euxanthe, and is preoccupied in Crustacea (Leach, 1813). See also Godartius.

92. ANTIGONIS.*

1861. Feld., Neues Lep. 21 : *Pharsalia*. Sole species, and therefore type. Used subsequently by Herrich-Schaeffer and Kirby in the same sense.

This name is preoccupied in several ways by the following names: Antigonus (Hübn., Lep. 1816), Antigona (Schum., Moll. 1817), and Antigonia (Lowe, Fishes, 1844). Kirby has proposed the name Lincoya (q. v.) for this group.

93. ANTIGONUS.

1816. Hübn., Verz. 108 : **Nearchus** (ustus), Erosus.
1870. Butl., Ent. Monthl. Mag. vii. 98 : designates Nearchus as the type. See also Chætoneura.

94. ANTIRRHEA.

1822–26. Hübn., Exot. Schmett. ii : **Archæa**. Sole species, and therefore type.
1844. Doubl., List Br. Mus. 121 : Archæa, Philoctetes.
1851. Westw., Gen. Diurn. Lep. 365 : uses it in the same sense.
1868. Butl., Ent. Monthl. Mag. iv. 195; and Cat. Satyr. 107 : gives Philoctetes as type, but of course erroneously.

He afterwards founded the genus Anchyphlebia upon Archæa, because Hübner's genus was not characterized; but see remarks under Anchyphlebia.

1871. Kirb., Syn. Cat. 38 : uses the genus in its proper sense.

95. Apatura.

1807. Fabr., Ill. Mag. vi. 280: Iris, **Bolina**, Alimena.

In 1806, Hübner (Tent.) selected Iris as type of Potamis; consequently Apatura must be restricted to the other two, which are congeneric, and Bolina may be taken as the type. This, however, is not in accordance with subsequent usage, as will be seen by the following:—

1815. Leach, Edinb. Encycl. 718: gives Iris only.
1816. Ochs., Schmett. Eur. iii. 19: gives Iris and Ilia; but he was restricted to these from the nature of the case.
1816. Hübn., Verz. 35: uses it for Bisaltide and a number of others, none of which have any thing to do with the Fabrician members of the genus.
1831. Curtis, Brit. Ent., pl. 338: designates Iris as type.
1832. Dup., Pap. France, Diurn. Suppl. 402: uses it for Iris and Ilia.
1833–4. Boisd.-LeC., Lép. Am. Sept. 206: refer Idyia (Clyton) and celtis to it.
1837. Sodoffsk., Bull. Mosc. x. 81: proposes to spell it Apaturia.
1840. Westw., Gen. Syn. 87: specifies Iris as type.
1844. Doubl., List Br. Mus. 108: refers to it Iris, Ilia, and Clyton.
1850. Westw., Gen. Diurn. Lep. 302: regards Iris and Ilia as types.
1861. Feld., Neues Lep. 36: divides the group into six sections, to the first of which he gives the name of Apatura *par excellence*, with Iris, Ilia, and Namouna (Ambika) as species.
1871. Kirb., Syn. Cat. 259: uses it in the extended Felderian sense, arranging the species in the same order.
1872. Crotch, Cist. Ent. i. 66: says Iris is type, on account of Ochsenheimer's limitation, overlooking the work of his own countryman, Leach.

This result is from want of familiarity with Hübner's Tentamen. See also Esoptria, Æola, Hypolimnas, Diadema, and Potamis.

96. Apaturia.*

1837. Sodoffsk., Bull. Mosc. x. 81: proposes this name as an etymological correction for Apatura (q. v.).

97. Apaturina.

1865. Herr.-Schaeff., Prodr. i. 75: **Erminea**. Sole species, and therefore type.

98. APAUSTUS.

1816. Hübn., Verz. 113: **Menes**. Sole species, and therefore type.
Butler and Kirby use it subsequently in the same sense.

99. APHACITIS.

1816. Hübn., Verz. 19: Lusca, **Lucinda** (Dyndima).
Lusca, though Hübner's species, was not published until after his death, and hence we must take Lucinda as the type. See Nelone.

100. APHANTOPUS.*

1853. Wallengr., Lep. Scand. Rhop. 30: *Hyperanthus*. Sole species, and therefore type.
Falls before Hipparchia (q. v.).

101. APHNÆUS.

1816. Hübn., Verz. 81: Vulcanus, **Orcas**.
1847. Doubl., List Brit. Mus. 25: employs it for a number of species, including both of Hübner's.
1858. Horsf.-Moore, Cat. Lep. E. Ind. Co. i. 37: employ it for both Hübner's species and others.
Herrich-Schaeffer, Butler, and Kirby also use it for both of Hübner's species with others. Orcas may be taken as type.

102. APHRISSA.

1873. Butl., Lep. Exot. 155: **Statira**. Sole species and designated type.

103. APHRODITE.*

1816. Hübn., Verz. 95: Evippe, Danæ (Eborea).
This name is preoccupied by Aphrodita (Linn., Worms, 1735). See Callosune.

104. APODEMIA.*

1865. Feld., Reise Novara, 302: Mormo, virgulti (Sonorensis).
This name also is preoccupied by Apodemus (Kaup, Mammals, 1825).

105. APORIA.

1816. Hübn., Verz. 90: **cratægi**. Sole species, and therefore type.
It has frequently been used (Stephens, Wallengren, Staudinger, Westwood) in the same sense. See Leuconea and Pieris.

106. APOSTRAPHIA.

1816. Hübn., Verz. 13: Ricini, Bellona (Brassolis), **Charithonia** (Charitonia).
Bellona (not a Heliconian at all) was placed here by error. Charithonia may be taken as type.

107. APPIAS.

1816. Hübn., Verz. 91: **Zelmira**, Achine.

As Achine is needed as type of Anthopsyche, Zelmira may be considered the type of this group.

108. APROTOPOS.* [Aprotopus in Index.]

1871. Kirb., Syn. Cat. 19: Ædesia, Ceto, Melantho, Pytho

Ædesia being the necessary type of Xanthocleis, this name must fall, unless one of the other species should prove generically distinct.

109. ARASCHNIA.

1816. Hübn., Verz. 37: **Levana** (Levana, Prorsa). Sole species, and therefore type.

110. ARCAS.*

1832–33. Swains., Zoöl. Ill. ii. 88: *imperialis*. Sole species, and therefore type.

Must this name fall before Evenus? (q. v.)

111. ARCHON.

1822. Hübn., Syst.-Alph. Verz.: Machaon (Machaon, Sphyrus), Medisicaste, Mnemosyne, Phœbus, Podalirius (Podalyrius), Polyxena, Rumina, **Apollinus** (Thia).

Apollinus may be taken as the type. See Doritis.

112. ARCHONIAS.

1825. Hübn., Zutr. iii. 19: **Tereas** (Marcias). Sole species, and therefore type. See Euterpe.

113. ARESTA.

1820. Dalm. in Billb., Enum. Ins. 79: Amestris, Idalia, Ariadne, Asterie, Cloantha, **Laomedia**.

Laomedia may be selected as the type of this genus.

114. ARGE.*

1816. Hübn., Verz. 60: occitanica (Psyche), Arge (Amphitrite), Thetis (Inis), Russiæ (Clotho), Lachesis, Galathea.

This generic term has been used largely by subsequent authors (Boisduval, Duponchel, Stephens, Doubleday, Westwood, Rambur, etc.), always in nearly the same sense; and Butler and Grote even cite occitanica (Syllius Butl., Psyche Grote) as the type; but the name cannot stand: first, because founded upon a name used for one of the original species; and, second, because preoccupied in Hymenoptera (Schrank, 1801). See Agapetes and Melanargia.

115. Argus.*

1764. Geoffr., Hist. des Ins. ii. 61: employs the term *Les Argus* at the head of a division of blues; but it can have no weight as a generic name, because it is used simply as a French word, as *Les Estropiés* is for the next division, of skippers.

1777. Scop., Introd. 432: employs it for more than fifty species, having no common and distinctive structural bond; they are divided into four sections, all of them almost equally heterogeneous in composition, each, excepting the last, containing members of every family of butterflies excepting the Urbicolæ. The name must therefore be dropped, and not be employed again in any sense. [The species Argus was not included in the genus by Scopoli.]

Boisduval also uses it in his Species général, but is not followed in this use by many other authors.

1816. Lam., Hist. Nat. An. sans Vert. iv. 21: employs it for Argyrognomon (vulgaris), Corydon, and others. One of the synonymes of Argyroguomon is Argus, so that if it be considered that the name was founded anew in this instance, it must be dropped, because based on a specific name.

1832. Dup., Pap. de France, Diurn. Suppl. 388: Battus and many others.

1832. Boisd., Icon. 49: employs it for the blues of Europe, appending his own name as authority!

1832. Ib., Voy. Astrol. 90: Cleotas (Poeta).

1833. Ib., Nouv. Ann. Mus. Hist. Nat. ii. 171: Lysimon.

1833–34. Boisd.-LeC., Lép. Amér. Sept. 113: Hanno (Filenus), etc.

1838–39. Krause, Faun. Thur. 60: uses it for Blues and Coppers.

1872. Scudd., Syst. Rev. 6: wrongly attempts to revive the name, calling Eurydice, one of Scopoli's species, the type.

116. Argynnina.

1867. Butl., Ann. Mag. Nat. Hist. [3] xix. 165: **Hobartia**, Lathoniella.

1868. Ib., Ent. Monthl. Mag. iv. 196: Hobartia specified as type.

117. Argynnis.

1807. Fabr., Ill. Mag. vi. 283: I. Paphia, Maia (Cynara), Laodice (Cethosia), **Aglaja**; II. Liriope, Tharos (Morpheus), Hermes.

1810. Latr., Consid. 440: specifies Paphia and Cinxia as types; but Paphia, the only one of these specified by Fabricius, cannot be the type, because already the type of Dryas (q.v.).
1815. Leach, Edinb. Encycl. 717: restricts it to the first of the Fabrician sections.
1815. Oken, Lehrb. i. 734: gives it the same restriction, as have all subsequent authors.
1816. Dalm., Vetensk. Acad. Handl. xxxvii. 57, 66: I. Paphia, Aglaja, Adippe, Niobe, Lathonia (Latonia); II. Aphirape, Selene, Euphrosyne, Amathusia, lapponica (Freja), Pales, Dia, Chariclea (Carichlea), Frigga, Ino, Thore. Adippe specified as type, but of course erroneously.
1816. Hübn., Verz. 30: Aphirape and its allies.
1820. Oken, Lehrb. f. Schulen, 790: Aglaja only.
1830. Curtis, Brit. Ent., pl. 290: specifies Aglaja as type.
1840. Westw., Gen. Syn. 88: wrongly specifies Paphia as type.
1872. Scudd., Syst. Rev. 24: specifies Aglaja as type.
1872. Crotch, Cist. Ent. i. 66: again specifies Paphia as type. See also Argyronome.

118. ARGYREA.*

1820. Billb., Enum. Ins. 77: vanillæ, Lathonia, Niobe, Adippe, Aglaia (Aglaja), Paphia, Maia (Cynara), Niphe, Phalanta [?] (Pharantha), Aphirape, Selene, Euphrosyne, Pales, Gersenii, Ino, Thore, Amathusia, lapponica (Freja), Frigga.

This term is preoccupied by Argyreus (Scop., Lep. 1777) and Argyria (Hübn., Lep. 1816).

119. ARGYREUS.*

1777. Scop., Introd. 431: Niphe and twenty-six others in two sections, the former of which is divided into five, and the latter into three subsections; but they are all brought together in such a confused manner, and formed of such utterly incongruous material, even to what must have been the sense of the naturalists of his own day, that the genus must fall into merited oblivion. Subsection c of section A contains, for example, the following species among others: Rumina [Papilionides], vanillæ [Nymphales], and Cupido [Rurales].

120. ARGYRONOME.

1816. Hübn., Verz. 32: Lampetia, Phalanta (Columbina), **Laodice**, Paphia, Maia (Pandora), Hostilia (Orthosia).

1850. Steph., Cat. Br. Lep. 13, 258: uses it for Paphia, Aphrodite, and Cybele; but Paphia, the only one of Hübner's species, is the type of Dryas.

Should Laodice prove generically distinct from the species of the genera Dryas and Argynnis, this name may be reserved for it; otherwise it will fall.

121. ARGYROPHÉNGA.

1848. Doubl., List Br. Mus., App. 31: **antipodum**. Sole species, and therefore type.

Used in this sense by subsequent writers.

122. ARGYROPHORUS.

1852. Blanch., Gay's Chili, vii. 30: **argenteus**. Sole species, and therefore type.

So used by Butler.

123. ARHOPALA.

1832. Boisd., Voy. Astrol. 75: **Phryxus**. Sole species, and therefore type.

124. ARIADNE,[*]

1829. Horsf., Descr. Cat. Lep. E. Ind. Co., expl. pl.: *Ariadne* (Coryta). Sole species, and therefore type.

1833. Boisd., Ann. Mus. Nat. Hist. 201: specifies Merione and Coryta as types.

But, being founded upon the name of its original species, the generic name must be dropped, and not be brought again into use. It is, moreover, preoccupied in Arachnids (Sav. 1825). See Ergolis.

125. ARICORIS.

1851. Boisd. in Westw., Gen. Diurn. Lep. 449: Cepha (Epitus), **Tisiphone**, Tutana, Constantius, Theanus.

1868. Bates, Journ. Linn. Soc. Lond. ix. 456: employs it for Cepha (Epitus) and others.

1871. Kirb., Syn. Cat. 332: uses it in the same sense as Bates.

Cepha, however, cannot be taken as type, for in 1856, through Pándemos, this became the type of Boisduval's genus Orimba; nor Theanus, for this is the type of Trichonis (1865); Constantius is too far removed from the others to be looked upon as at all typical, so that the choice remains, notwithstanding the action of Bates and Kirby, between Tisiphone and Tutana. We propose that Tisiphone be considered the type, since it is illustrated by Westwood.

126. ARISBA.*
1847. Doubl., List Br. Mus. 11: *Agacles.* Sole species, and therefore type.
   Preoccupied by Arisbe (Hübn., Lep. 1816).

127. ARISBE.
1816. Hübn., Verz. 89: **Leonidas** (similis), Panope.
   Leonidas may be taken as the type.

128. ARMANDIA.
1871. Blanch., Comptes Rend. lxxii. 809: **Thaidina.** Sole species, and therefore type. See Bhutanitis.

129. AROTES.*
1851. Boisd. in Westw., Gen. Diurn. Lep. 450: given by Westwood as a MS. synonyme of Setabis (q. v.).

130. ARPIDEA.*
1837. Dunc., For. Butt. 180: *Chorinæa.* Sole species, and therefore type.
   This name must fall before Cærois (q. v.), and not be used again. See also Hames.

131. ARTEUROTIA.
1872. Butl.-Druce, Cist. Ent. i. 112: **tractipennis.** Sole species and designated type.

132. ARTIPE.*
1870. Boisd., Lép. Guat. 14: *Eryx* (Amyntor) specified as type.
   But the generic name is preoccupied by Artipus (Schönh., Col. 1826). See Deudorix.

133. ASCANIDES.
1837. Gey. in Hübn., Zutr. v. 32: **Triopas.** Sole species, and therefore type.

134. ASCIA.
1777. Scop., Introd. 434: cratægi, napi, sinapis, **Monuste**, Polybe.
   With the exception of the last species, which belongs to the Rurales, the genus is comparatively homogeneous, — the only one of Scopoli's of which this can be said, — and it should therefore be retained for one of the groups included in it. See Mylothris.

1816. Billb., Enum. Ins. 79: gives this name to a number of species belonging to the Nymphales, using Scopoli's name at the same time as the author!

1872. Scudd., Syst. Rev. 40: restricts the name to Monuste, and correctly; for Aporia removed crataegi from this genus in 1816, and Leptidia, sinapis in 1820. Polybe belongs to a different family, and must be passed over; and napi cannot be used, as it is needed for Pieris (q. v.).

135. Asterope.

1816. Hübn., Verz. 66: Amulia, **Sapphira** (Sapphyra), Theanus (Theane).
    Sapphira may be taken as type. See Callithea.

136. Astictopterus.

1860. Feld., Wien. Ent. Monatschr. iv. 401: **Jama**, Sindu.
1870. Butl., Ent. Monthl. Mag. vii. 95: specifies Jama as type.

137. Astraptes.

1816. Hübn., Verz. 103: Corytas (Corytus), Pervivax, Narcosius, Apastus, Enotrus, Creteus, Mercatus (Fulgurator), **Aulestes**, Amyntas (lividus).
    Aulestes may be taken as the type.

138. Astycus.

1825. Hübn., Catal. Franck, 85: Peleus, Mercatus (Fulgerator), vitreus, Simplicius, Proteus? Evadnes, Exadeus? Thraso, erosus, Tryxus, orbifera (orbifer?), oileus, Carthami, alceæ (malvæ), Morpheus (Steropes), Crinisus, **Augias**, Actæon, Thaumas (linea), Arsalte (Menalcas), Talaus, Phyllus, and a MS. species.
1869. Herr.-Schaeff., Prodr. iii. 45, 54: suggests its employment, but does not indicate its membership.
    Peleus, Mercatus, Vitreus, Proteus, Thraso, Morpheus, Thaumas, and Arsalte are specified as types of other genera. Augias may be taken as the type of this, since it is the only one of the true Astyci[*] not already confined to a generic name which will hold.

139. Atalopedes.

1872. Scudd., Syst. Rev. 57: **Huron**, campestris. Huron specified as type.

140. Atella.

1847. Doubl., Gen. Diurn. Lep., pl. 22: **Phalanta** (Eurytis). Sole species, and therefore type.

---

[*] Cf. Bull. Buff. Soc. Nat. Sc. i. 195.

1848. Ib., ib. 165: Phalanta (Eurytis) and five others in three sections.

As the figured species appeared six months before the text, it must be considered the type of the genus, as indeed it has been virtually treated by subsequent writers. The name is rather close to Atela (Dej., Col. 1833). See Phalanta and Messaras.

141. ATERICA.

1833. Boisd., Ann. Mus. Hist. Nat. 195: **Rabena.** Sole species, and therefore type.

Used in same sense by all subsequent writers.

142. ATHENA.

1816. Hübn., Verz. 36: **Peleus** (Thetis). Sole species, and therefore type.

According to Kirby (Syn. Cat. 220), this name is preoccupied, but he does not state where. See also Petreus.

143. ATHESIS.

1847. Doubl., Gen. Diurn. Lep. i. 109: **Clearista.** Sole species, and therefore type.

144. ATHIS.*

1816. Hübn., Verz. 101: *Palatinus.* It is not a butterfly.

145. ATHYMA.

1850. Westw., Gen. Diurn. Lep. 272: **Leucothoe,** Aceris (Eurynome), Heliodora (Helicopis), Sulpitia (Strophia), opalina, Sankara, Larymna, Venilia, Saclava, Vikasi, Nefte, Inara, Melaleuca, Brebissonii.

1861. Feld., Neues Lep. 32: divides the group into two sections, the first containing Leucothoe and Larymna, the second Nefte, Inara, and Sulpitia (Strophia).

1865. Herr.-Schaeff., Prodr. i. 67: uses it for Leucothoe and Larymna only.

Leucothoe as the older species may be considered as the type.

146. ATHYRTIS.

1862. Feld., Wien. Ent. Monatschr. vi. 413: **Mechanitis.** Sole species, and therefore type.

Used in same way by Herrich-Schaeffer and Kirby.

147. ATLIDES.

1816. Hübn., Verz. 80: **Halesus** (Halesus, Dolichus), Polybe (Atys, Scamander).
1869. Butl., Cat. Fabr. Lep. 197: uses it in the same sense.
   Halesus may be considered as the type.

148. ATROPHANEURA.

1864. Reak., Proc. Ent. Soc. Philad. iii. 446: **Semperi** (Erythrosoma). Sole species, and therefore type.

149. ATRYTONE.

1872. Scudd., Syst. Rev. 56: **Iowa**, Logan, conspicua, Zabulon. Iowa specified as type.

150. AUGIADES.

1816. Hübn., Verz. 112: crinisus, Arcalaus, comma, **sylvanus**, Helirius, Euribates.
1850. Steph., Cat. Brit. Lep. 23, 263: uses it for sylvanus, comma, Vitellius.
1858. Kirb., List Brit. Rhop.: uses it for Vitellius only, but this is not congeneric with either sylvanus or comma.
1870. Butl., Ent. Monthl. Mag. vii. 58: overlooking the restriction of Stephens, calls crinisus the type.
1872. Scudd., Syst. Rev. 58: designates sylvanus as the type.
   See Erynnis.

151. AULOCERA.

1867. Butl., Ent. Monthl. Mag. iv. 121: **Brahminus**, Saraswati, Padma (Padma, Avatara), Scylla.
1868. Ib., Cat. Sat. 49: specifies Brahminus as type.
   Is this name too near Autocera (Melly, Col. 1857)?

152. AUROTIS.

1816. Dalm., Vetensk. Acad. Handl. xxxvii. 63, 90: **quercus**, betulæ, pruni, w. album, ilicis.
   It is given as a subgenus of Zephyrus, of which betulæ is type.
1863. Kirb., List Eur. Butt. 8: roboris (Evippus). [See also p. 293.]
   The last three of Dalman's species belonging to Thecla (q. v.) after the foundation of Zephyrus, quercus must be taken as the type of Aurotis, if it is generically distinct from betulæ; if not, Aurotis falls.

153. AUSTROMYRINA.*

1865. Feld., Reise Novara, 260: Evagoras, Ictenus (Schraderi).
   This name falls before Jalmenus.

154. Autochton.

1823. Hübn., Zutr. ii. 13: **Itylus**. Sole species, and therefore type.

155. Autodea.*

1850. Boisd. MS. in Westw., Gen. Diurn. Lep. 253: stated by Westwood to be synonymous with Hübner's Lucinia, as used in the Genera of Diurnal Lepidoptera. Of course it died at its birth.

156. Autonema.*

1850. Boisd. in Westw., Gen. Diurn. Lep. 266: Westwood states that this is a MS. synonyme of Prothoe (q. v.).

It is nowhere else referred to.

157. Axiocerses.

1816. Hübn., Verz. 72: **Perion**. Sole species, and therefore type.
1871. Kirb., Syn. Cat. 337 [Axiocerces]: uses it for Zeuxo and many others, including Perion. But see his Preface.

158. Bæotis.

1816. Hübn., Verz. 21: **Hisbon** (Hisbæna), Eumeus (Uranis).
1847. Doubl., List Brit. Mus. 11: uses it for Hisbon and others, not including Eumeus.
1851. Westw., Gen. Diurn. Lep. 451: divides the group into two sections, and in the second places Hisbon. Eumeus is not given.
1867. Bates, Journ. Linn. Soc. Lond. ix. 444 [Bœotis]: uses it for Hisbon and three others. Hisbon therefore becomes the type.

159. Barbarus.*

1872. Crotch, Cist. Ent. i. 60: refers this name, in a generic sense, to Linné, but remarks that it has not been accepted, because heterogeneous.

It does not seem to me to have ever been used, even by Linné, in a generic sense.

160. Barbicornis.

1823. God., Encycl. méth. ix. 705: **basilis**. Sole species, and therefore type.

Used in same sense by Westwood, Bates, and Kirby. Is it a butterfly? See Chroma.

161. BASILARCHIA.

1872. Scudd., Syst. Rev. 8: Archippus (Disippe), **Astyanax**, Artemis (Arthemis). Type specified as Astyanax.
See also Callianira.

162. BASSARIS.*

1816–21. Hübn., Exot. Schmett. ii.: *Itea*. Sole species, and therefore type.
1821. Hübn., Index, 4: Itea.
The name falls before Vanessa (q. v.). See also Ammiralis and Pyrameis.

163. BATESIA.

1862. Feld., Wien. Ent. Monatschr. vi. 112: **Hypochlora**. Sole species, and therefore type. See also Pandora.

164. BATTUS.*

1777. Scop., Introd. 433: Polydamas and a great number of wholly unrelated species, divided into six sections.

The utterly heterogeneous nature of this group may be shown by noticing a few of the species from the first section, such as Polydamas [Papilionides], Antiopa [Nymphales], Tespis [Rurales], and malvæ [Urbicolæ]. Of course the name must be dropped in perpetuity; moreover, Scopoli included in this group a species which he called Argus, but which the Therisianer called Battus, and the name should drop from this cause; nevertheless:—

1858. Ramb., Cat. Lep. Andal. 85: uses it for Sao.

This was not even one of the many original species, although (almost necessarily!) closely allied to some of them.

165. BELENOIS.

1816. Hübn., Verz. 92: **Calypso**. Sole species, and therefore type, as specified by Butler (Cist. Ent. i. 37, 50).

166. BHUTANITIS.*

1873. Atk., Proc. Zoöl. Soc. Lond. 570: *Lidderdali*. Sole species, and therefore type.
Falls, according to Kirby (in litt.), before Armandia.

167. BIA.

1816. Hübn., Verz. 51: **Actorion** (Actoriæna). Sole species, and therefore type.
Used in same sense by Westwood, Herrich-Schaeffer, and Kirby.

168. BIBLIS.*

1807. Fabr., Ill. Mag. vi. 281: Biblis, Leucothoe, Nauplia, Neærea.
1819. God., Encycl. méth. 325: employs it for Biblis (Thadana) and others.
1836. Boisd., Spec. gén., pl. 5 B.: uses it for Aganisa, closely allied to Biblis.

Falls from having been named after one of the species on which it is founded. See Zonaga.

169. BICYCLUS.

1871. Kirb., Syn. Cat. 47: **Hewitsonii**, Iccius, Italus, Zinebi.
1873. Ib., Zoöl. Rec. for 1871, 363: specifies Hewitsonii as type.

Correctly, since it was the type of Idiomorphus (q. v.), which this supplants.

170. BITHYS.

1816. Hübn., Verz. 75: Erix (Tyrrhenus), Cupentus (Cubentus), Cethegus, Vesulus, **Strephon** (Sicheus, Strephon), Lydus, Tephraeus, Leucophaeus, Sphinx, quercus.
1850. Steph., Cat. Brit. Lep. 17: uses it for quercus.
1858. Kirby, List Brit. Rhop.: uses it in the same way.
1869. Butl., Cat. Fabr. Lep. 186: employs it for Strephon, Cyllarus, Agrippa, and Dindymus.

The usage of Stephens and Kirby is indefensible, as quercus must belong to Aurotis. In accordance with Butler's action, Strephon may be taken as the type.

171. BLETOGONA.

1867. Feld., Reise Novara, 465: **Mycalesis**. Sole species, and therefore type, as specified by Butler (Cat. Sat.).

172. BRACHYCNEME.*

1869. Herr.-Schaeff., Prodr. iii. 52.

No species are cited, and the name is preoccupied by Brachycnemis (Schönh., Col. 1844).

173. BRACHYGLENIS.*

1862. Feld., Wien. Ent. Monatschr. vi. 73: *Esthema*. Sole species, and therefore type.

According to Felder (ib. 285), the name is preoccupied (Brachyglene, Lep., Herr.-Schaeff.*). See Tmetoglene.

---

* I have not been able to find any such generic name in the works of Herrich-Schaeffer; nor is Mr. A. R. Grote, to whom I referred the question, acquainted with it. It is not given in Marschall's Nomenclator Zoölogicus.

174. BRANGAS.

1816. Hübn., Verz. 80: **Caranus** (Pelops, Caranus), Didymaon (Dydimaon), Syncellus, Bitias.
1869. Butl., Cat. Fabr. Lep. 196: uses it for Thales, Caranus, Didymaon.

Caranus may be taken as type.

175. BRASSOLIS.

1807. Fabr., Ill. Mag. vi. 282: **Sophoræ**, cassiæ, Obrinus.
1816. Hübn., Verz. 50: uses it for Darius (Anaxerete), and others, including Sophoræ and cassiæ.
1823. God., Encycl. méth. 456: employs it for Sophoræ and the allied Astyra.
1851. Westw., Gen. Diurn. Lep. 341: indicates Sophoræ as type.
1871. Crotch, Cist. Ent. i. 66: does the same.

176. BRENTHIS.

1816. Hübn., Verz. 30: **Hecate**, Ino (Dictynna), Thore, Daphne, Claudia.
1861. Feld., Neues Lep. 10: divides the group in two sections, specifying no species for the first, and for the second Pales and Cytheris.
1865. Herr.-Schaeff., Prodr. i. 73: gives Cytheris (Siga, Cytheris) and others, including none of Hübner's, all but the last of which are placed in Arygnnis.
1872. Scudd., Syst. Rev. 24: indicates Hecate as type.

177. BRONTIADES.

1816. Hübn., Verz. 113: **Procas**, Gentius, Arsalte (Menalcas) Petrus.
1870. Butl., Ent. Monthl. Mag. vii. 94: designates Procas as type.

178. BUTLERIA.

1871. Kirb., Syn. Cat. 624: Polyspilus, **exornatus**, Agathocles, Cypselus, Caicus, Cœnides, dimidiatus, Polycrates, Epiphaneus, Hesperioides, auricipennis, bisexguttatus.
1873. Ib., Zoöl. Rec. for 1871, 365: specifies exornatus as type.

See also Carterocephalus.

179. BYBLIA.

1816. Hübn., Verz. 28: **Ilithya**. Sole species, and therefore type.

See also Hypanis.

180. CABIRUS.

1816. Hübn., Verz. 102: Linus, **Julettus.**
Linus is not a butterfly, and Julettus may be taken as the type.

181. CÆCINA.

1868. Hewits., Hundr. Hesp. 55: **Calathana**, compusa.
Calathana may be considered as the type.

182. CÆROIS.

1816. Hübn., Verz. 56: **Chorinæus** (Arcesilae). Sole species, and therefore type, as stated by Butler (Cat. Sat. 1).
1851. Westw., Gen. Diurn. Lep. 366: the same.
1865. Herr.-Schaeff., Prodr. i. 63 [Cærous]: the same.
See Arpidea and Hames.

183. CALAIDES.

1816. Hübn., Verz. 86: **Androgeos** (Polycaon, Androgeus, Piranthus), Menatius.
Androgeos may be taken as the type.

184. CALAIS.*

1836. Boisd., Spec. gén. 584: given as a MS. synonyme of Idmais (q. v.).
It has never been used, and of course falls; moreover, it is the name of one of the species upon which it was proposed to found it.

185. CALEPHELIS.

1869. Grote-Rob., Trans. Am. Ent. Soc. ii. 310: **Cæneus** (Cænius), borealis. Type specified as Cæneus.

186. CALIGO.

1816. Hübn., Verz. 51: Teucer (Teucra), Idomeneus (Idomenea), **Eurylochus** (Euriloche), Ilioneus (Ilionea).
1844. Doubl., List Br. Mus. 117: uses it for four species, allied to those of Hübner, but including none of them.
1851. Westw., Gen. Diurn. Lep. 340: employs it for Hübner's species and others, specifying Teucer and Eurylochus as the types.
1864. Herr.-Schaeff., Prodr. i. 55: uses it similarly.
1870. Boisd., Lép. Guat. 54: uses it for species placed by Westwood and Kirby in the allied genus Opsiphanes.
1871. Kirb., Syn. Cat. 127: employs it in the Westwoodian sense.
Eurylochus may be taken as type. The name is very close to Caligus (Müll., Crust. 1785). See Ærodes.

187. CALINAGA.

1858. Moore, Cat. Lep. East Ind. Co. i. 162: **Buddha.** Sole species, and therefore type.

188. CALISTO.

1823. Hübn., Zutr. ii. 16: **Herophile.** Sole species, and therefore type. It has been used in this sense by subsequent writers; but
1868. Butl., Ent. Monthl. Mag. iv. 194; and Cat. Sat. 97: specifies Zangis as type, of course erroneously.

189. CALLEREBIA.

1867. Butl., Ann. Mag. Nat. Hist. [3], xx. 217: **Scanda** (Scanda, Armanda), Nirmala. Scanda is specified as type, as also subsequently (Cat. Sat.; Ent. Monthl. Mag. iv. 194).

190. CALLIANIRA.*

1816. Hübn., Verz. 38: *Astyanax* (Ephestiæna). Sole species, and therefore type.
1844. Doubl., List Br. Mus. 91: gives Eurota with a query, but this is far removed from Hübner's type.
1850. Westw., Gen. Diurn. Lep. 251: refers the generic name to Boisduval! and places in it Alcmena, Eurota, and others, with equal error.

The name is, however, preoccupied in Mollusks (Pér.-Les. 1810). See also Basilarchia.

191. CALLICORE.

1816. Hübn., Verz. 41: **Codomannus** (Astarte), Clymena.
1849. Doubl., Gen. Diurn. Lep. 237: Clymena and eleven others.
1861. Feld., Neues Lep. 20: cites no species, though Pandama and Bacchis are said to belong here, but erroneously [see Cyclogramma].
1869. Butl., Cat. Fabr. Lep. 64: gives Clymena (Janeira) only.
1871. Kirb., Syn. Cat. 207: also gives Clymena and a dozen other species, omitting Codomannus, which is not congeneric.

Notwithstanding the limitation of Doubleday, Butler, and Kirby, Clymena cannot be taken as type, since Billberg has earlier (Enum. Ins. 1820) selected this as the type of Diæthria (q. v.); and hence Codomannus must be the type. See also Catagramma.

192. CALLIDRYAS.

1829–30. Boisd.-LeC., Lép. Am. Sept. 73 : **Eubule.** Sole species, and therefore type.
1832. Boisd. in Poey, Cent. Lép. Cuba, i.: Orbis.
1832. Ib., Voy. Astrol. 62 : Pomona, Crocale (Eudeer).
1836. Ib., Spec. gén. 605 : gives twenty-six species, including all the above, placing them in three groups.
1870. Butl., Cist. Ent. i. 36, 46 ; and Lep. Exot. 155 : designates Eubule as type.

193. CALLIDULA.*

1816. Hübn., Verz. 66 : Evander (Evandra), Petavius (Petavia), Pyramus (Pyrame).

The first two species are not butterflies, and the genus may therefore be referred to the heterocerous Lepidoptera.

194. CALLIMORMUS.

1872. Scudd., Syst. Rev. 53 : **juventus.** Sole species and designated type.

195. CALLIONA.

1868. Bates, Journ. Linn. Soc. Lond. ix. 447 : **Irene,** Latona, Siaka.
Irene may be considered as the type.

196. CALLIPAREUS.

1872. Scudd., Syst. Rev. 30 : **Melinus.** Sole species and designated type.

197. CALLITÆNIA.*

1861. Feld., Neues Lep. 50 : no species (but an unnamed MS. one) cited.
1865. Herr.-Schaeff., Prodr. i. 82 : refers Doris (Feld., Wien. Ent. Monatschr. 1860, 107) to this.

The name is, however, preoccupied by Calotænia ("Scr. Callitænia," Agass. Nomencl. Zoöl.), a genus of Lepidoptera (Steph. 1829). See Mesotænia.

198. CALLITÆRA.*

1868. Butl., Cat. Sat. 101 : Menander (Menander, Andromeda), Pireta (Aurora), Andromeda (Esmeralda), Philis (Harpalyce).

This generic name falls before Cithærias (q. v.).

199. CALLITHEA.*

1836. Boisd., Spec. gén., pl. 6 B.: Sapphira. Sole species, and therefore type.

Subsequently used by Westwood, Felder, and Kirby. The name, however, must be dropped, because based upon one of the names of the species upon which it is founded. See also Asterope.

200. CALLITHOMIA.

1862. Bates, Linn. Trans. xxiii. 522: **Alexirrhoe**, Zeuxippe, Thornax.

Alexirrhoe may be taken as the type.

201. CALLIZONA.*

1848. Doubl., Gen. Diurn. Lep., pl. 296: *Aceste.* Sole species, and therefore type.
1850. Westw., ib. 246: Aceste (Acesta).

The name falls before Tigridia (q. v.), since that genus was restricted to Aceste by Doubleday's own action in 1844. The name is also preoccupied by Callizonus (Schönh., Col. 1826).

202. CALLOPHRYS.

1820. Billb., Enum. Ins. 80: Vulcanus, **rubi**, and a MS. species.

Rubi may be taken as type.

203. CALLOSUNE.

1847. Doubl., Gen. Diurn. Lep. 57: subfasciatus (subfasciata), Evanthe, Eucharis, Evarne, **Danæ**, Eupompe, Achine (Antevippe, Achine), Antigone, Evippe, Omphale, Thogone, Etrida, Phlegetonia, Delphine, Eione, Daira, Evagore, Ephyia (Ephya), Liagore, Eulimene, Cebrene, Ocale (Omphale, by misprint), Ione.

Wallengr. (Rhop. Caffr. 10), in founding his genus Anthopsyche, which was in general originally synonymous with this, divides it into two sections. We have above (see Anthopsyche) proposed to restrict Anthopsyche to his first section, and for his second suggest the retention of Callosune, with Danæ for type. See also Aphrodite and Anthopsyche.

204. CALORNIS.

1820. Billb., Enum. Ins. 77: Euterpe, Susanna, Rosalia, **Thalia**.

The first two species belong to Boisduval's Nerias (1836): Rosalia is the type of Sais (Hübner, 1816), by Doubleday's action in 1844; hence Thalia must be taken as the type of this group. See Actinote.

205. CALOSPILA.*

1832. Gey. in Hübn., Zutr. iv. 28: *Parthaon* (Thermodoë). Sole species, and therefore type.

This name is used in the same sense by Doubleday and Westwood, and in a different sense by Bates; but the name is preoccupied by Calospilus (Hübn., Lep. 1816). See Polystichtis and Lemonias.

206. CALPODES.

1816. Hübn., Verz. 107: exclamationis (Forulus), **Ethlius**.
1870. Butl., Ent. Monthl. Mag. vii. 93: gives it as a section of Pamphila with Ethlius and others.
1872. Scudd., Syst. Rev. 61: designates Ethlius as type.

This name is written by Hübner twice as Calpodes and, including its Teutonic form, three times as Colpodes in the Verzeichniss. Colpodes would seem to be the more probably correct form, judging from the derivation of the word; but in that case it would be preoccupied, through Colpoda (Schrank, Polyg. 1803), and it would therefore be better to retain it as Calpodes.

207. CALYDNA.

1847. Doubl., List Br. Mus. 5: Meris, **Thersander**.
1851. Westw., Gen. Diurn. Lep. 436: employs it for Thersander and a few others.
1867. Bates, Journ. Linn. Soc. Lond. ix. 441: uses it for Thersander and many others.
1871. Kirb., Syn. Cat. 317: follows Bates.

Thersander must therefore be considered as the type.

208. CAMENA.*

1865. Hewits., Ill. Diurn. Lep. ii. 47: *Ctesia*. Sole species, and therefore type.
1868. Herr.-Schaeff., Prodr. iii. 21 [Camœna]: the same.

Preoccupied through Camœna (Baly, Col. 1862).

209. CANDALIDES.

1816. Hübn., Verz. 73: **xanthospilos**, Thetys (Phædrus).

Thetys having become the type of Curetis, (the same species being given by Hübner in two genera!) xanthospilos becomes the type of this.

210. CANOPUS.*

1861. Wallengr. in Feld., Neues Lep. 33: *Dædalus* (Meleagris). Sole species, and therefore type.

This name is preoccupied in Hemiptera (Fabr. 1803) and Polyps (Montf. 1808). See Hamanumida.

211. CAPILA.

1865. Moore, Proc. Zoöl. Soc. Lond. 785: **Jayadeva.** Sole species, and therefore type.

> This name can scarcely be considered too close to Capella, used in Mammals (Keys and Blas, 1850).

212. CAPRONA.

1857. Wallengr., Rhop. Caffr. 51: **Pillaana.** Sole species, and therefore type, as specified by Butler.

213. CAPYS.

1865. Hewits., Ill. Diurn. Lep. 58: **Alphæus.** Sole species, and therefore type. See Scoptes.

214. CARCHARODUS.*

1816. Hübn., Verz. 110: lavateræ (lavatheræ), altheæ, alceæ (malvæ).

> This is subsequently used by Westwood, Stephens, and Kirby, but it falls before Urbanus. See also Spilothyrus.

215. CARIA.

1823. Hübn., Zutr. ii. 14: **Argiope** (Colubris). Sole species, and therefore type.

> Used by Erichson (Schomb. Reise, 1848) in a similar sense.

216. CARTEA.

1871. Kirb., Syn. Cat. 308: **Vitula,** Tapajona.
1873. Ib., Zoöl. Rec. for 1871, 364: designates Vitula as the type.

> Correctly, since it was the type of Orestias, which this name was intended to supplant.

217. CARTEROCEPHALUS.*

1852. Led., Verh. zoöl.-bot. Gesellsch. Wien, ii. 26, 49: *Palæmon* (Paniscus), Sylvius, argyrostigma.

> Although proposed by Lederer to supplant Steropes, preoccupied, none of the original species of Boisduval are cited, and the short diagnosis is taken from the species above mentioned.

1867. Snell., Vlind. Nederl. 83: gives Palæmon (Paniscus) as type.
1870. Butl., Ent. Monthl. Mag. vii. 96: specifies exornatus as type, but erroneously [see Butleria].

> The three species given by Felder are not congeneric with those originally specified by Boisduval under Steropes; they are mutually congeneric, however, and the name must fall before Pamphila, virtually limited in 1832 to this group.

218. Carystus.

1816. Hübn., Verz. 114: **Jolus**, Hylaspes, Phyllus, Abebalus.
1869. Butl., Cat. Fabr. Lep. 273: uses it for Phyllus and three others not mentioned by Hübner.
1870. Ib., Ent. Monthl. Mag. vii. 92: specifies Jolus as type.
1871. Kirb., Syn. Cat. 589: places all of Hübner's species and others in the group.

Phyllus and Jolus being strictly congeneric, Jolus can be taken as the type.

219. Castalia.*

1858. Boisd. in Horsf.-Moore, Cat. Lep. East Ind. Co. i. 199: *Dichroa*, Chandra.

It is used in same sense by Felder and others. Dichroa should be type, as the only species known to Boisduval. But the genus is preoccupied, both exactly, in Worms (Savig. 1817), in Mollusks (Lam. 1819), and in Coleoptera (Lap.-Gay, 1838); and also by Castalius, a genus of Lepidoptera (Hübn. 1816).

220. Castalius.

1816. Hübn., Verz. 70: Clyton, **Rosimon** (Naxus, Rosimon).
1869. Butl., Cat. Fabr. Lep. 162: uses it for Rosimon, which therefore becomes type.

221. Castnius.*

1816. Hübn., Verz. 102: Iphis (Juppiter, sic!), Pelasgus, Lycagus (Lucagus).

Pelasgus is a Castnian: the others belong to the Urbicolæ; but the name is preoccupied by Castnia (Fabr., Lep. 1807), of which it was probably intended as only an altered form.

222. Casyapa.

1871. Kirb., Syn. Cat. 576: **Corvus**, Cerinthus, Cariatus, Callixenus, Thrax, Thyrsis, Semamora, Divodasa, Chaya, Agna, Cinnara, Mangala.
1873. Ib., Zoöl. Rec. for 1871, 365: designates Corvus as type.

Correctly, since that was the type of Chætocneme, for which name this was substituted.

223. Catagramma.*

1836. Boisd., Spec. gén., pl. 5 B.: Pygas (Hydaspes). Sole species, and therefore type.

It is used in same sense by Doubleday, Felder, and Kirby; but Pygas is congeneric with Codomannus, and therefore it must fall before Callicore.

224. CATAGRAMMINA.

1867. Bates, Journ. Linn. Soc. ix. 411: **Tapaja.** Sole species, and therefore type.

225. CATARGYRIA.

1822–26. Hübn., Exot. Schmett. ii: Druryi, **Laurentia** (Seraphina), Laure (Laura).

1861. Feld., Neues Lep. 37: uses it for Cyane, Laurentia, Laure (Laura), and Druryi.

Laurentia may be taken as the type.

226. CATASTICTA.

1870. Butl., Cist. Ent. i. 34, 43: **Nimbice,** Semiramis, Bithys, Sebennica. Nimbice specified as type.

227. CATHÆMIA.

1816. Hübn., Verz. 92: **Cæneus** (Anthyparete), Isse, Ada, Agathina (xantholeuca), Belisama, Dorimene, Hirlanda.

1867. Herr.-Schaeff., Prodr. ii. 11: uses it for Belladonna and many others, including all of the above, excepting Ada, Agathina, and Hirlanda.

Cæneus may be taken as the type.

228. CATOCHRYSOPS.

1832. Boisd., Voy. Astrol. 87: Cyta, **Strabo,** Centaurus.

Strabo may be taken as the type.

229. CATONEPHELE.

1816. Hübn., Verz. 40: **Acontius** (Eupalemæna, Chione), Numilia, Cupavia.

1849. Doubl., Gen. Diurn. Lep. 222: uses it for Numilia (Micalia), Acontius (Medea), and Chromis.

Acontius may be considered the type.

230. CATOPHAGA.

1816. Hübn., Verz. 93: **Paulina,** Canidia (Gliciria), Cheiranthi, brassicæ, rapæ, napi (bryoniæ, napi).

Paulina may be taken as the type. See also Pieris.

231. CATOPSILIA.

1816. Hübn., Verz. 98: **Crocale,** Trite, Statira (Alcmeone), Pomona (Hilaria).

1871. Kirb., Syn. Cat 481: uses for Florella and a large number of others, including all of Hübner's.

1872. Scudd., Syst. Rev. 37: designates type as Crocale.
1873. Butl., Lep. Exot. 154: makes the same designation.

232. CATUNA.

1871. Kirb., Syn. Cat. 238: Crithea, **angustatum**, Opis, Cœnobita.
1873. Ib., Zööl Rec. for 1871, 360: designates angustatum as the type.

> Doubtless because it was supposed* to be the type of Felder's genus Euomma (preocc.) which this supplants. See also Jœra.

233. CAUDATI.*

1860. Koch, Stett. Ent. Zeit. xxi. 230: Daunus and a large number of tailed Papilionids.

> This group, being founded solely upon the presence of caudate appendages to the hind wings of Papilionides, would not have been excusable, scarcely tolerable, if it had been proposed in the middle of the last century; it is astonishing that it was allowed to appear in the respectable journal of Stettin; of course it must drop, even if the name were not preoccupied (Dum., Rep. 1806) or its form unobjectionable. It is also used by Swainson (Zoöl. Ill.) for a division of swallow-tails, but not in a generic sense.

234. CECROPS.*

1816. Hübn., Verz. 104: bipunctatus (Neis), Zarex.

> Preoccupied in Crustacea (Leach, 1818).

235. CECROPTERUS.

1869. Herr.-Schaeff., Prodr. iii. 45: no species are cited, but it is intended to supplant the preoccupied Cecrops.
1871. Kirb., Syn. Cat. 634: gives **Zarex**, Oryx, Phrynicus, thus putting Herrich-Schaeffer's suggestion into practice.

> Zarex may therefore be considered as the type.

236. CELÆNORRHINUS.

1816. Hübn., Verz. 106: Corbulo, Niso, **Eligius**, Cebrenus, Sergestus, Lucifer, Phæomelas.

> Eligius may be selected as the type. See Plesioneura.

237. CELŒNA.

1849. Boisd. in Doubl., Gen. Diurn. Lep. 214 [Celæna]: Doubleday gives this name as a MS. synonyme for Anartia (q. v.).

---

* But incorrectly; see Euomma.

1870. Boisd., Lép. Guat. 32: employs it for **Fatima**.

    This is one of the species included in it by Doubleday, and therefore may be taken as the type of Boisduval's Colœna; since the species is generically distinct from Jatrophæ, the type of Anartia, the genus will stand, but date from 1870.

238. Cepora.

1820. Dalm. in Billb., Enum. Ins. 76: Monuste (Heliades MS.), brassicœ, Canidia (Gliciria), rapæ, napi, **Nerissa** (Coronnis), Daplidice, cardamines, Eupheno, Eucharis, Glaucippe, and a number of MS. species.

    Nerissa may be taken as the type.

239. Ceratinia.*

1816. Hübn., Verz. 10: Eumelia, Lenea (Lenea, Melanida), *Nise*, (Neso), Ninonia.
1844. Doubl., List Br. Mus. 57: uses it for Nise and Lenea (Lenea, Melanida).
1847. Ib., Gen. Diurn. Lep. 127: employs it for Nise and many others, excluding Lenea.
1862. Bates, Linn. Trans. xxiii. 523: limits it again to seven species, of which the only one of Hübner's is Ninonia, which was not used by Doubleday in the first instance, although subsequently employed by him.
1870. Boisd., Lép. Guat. 32 [Ceratonia]: employs it for a number of species, including Ninonia (Barii, Ninonia).
1871. Kirb., Syn. Cat. 21: follows Bates.

    By Doubleday's restriction, however, Nise must be considered as the type. But the name is preoccupied through Ceratina (Latr., Hym. 1804).

240. Ceratrichia.

1869. Butl., Cat. Fabr. Lep. 274: **Nothus**, Phocion. Nothus designated as type.

241. Cethosia.

1807. Fabr., Ill. Mag. vi. 280: **Cydippe**, Biblis (Biblis, Penthesilea).
1809. Latr., Gen. Crust. et Ins. iv. 200: divides the group into two sections, thus: I. Juno, Julia (Alcionea); II. Cydippe, Biblis (Penthesilea); the second corresponding to the Fabrician idea.

1810. Ib., Consid. 440: designates Cydippe and Juno as types. Since Cydippe alone was mentioned by Fabricius, it becomes the type.
1820. Billb., Enum. Ins. 78: unreasonably changes the generic name to Eugramma (q. v.).

All subsequent authors have followed Latreille in the definition of the group.

1872. Crotch, Cist. Ent. i. 65: notices Cydippe as type, as above.
See Alazonia.

242. CHÆTOCNEME.*

1860. Feld., Sitzungsb. Acad. Wien. xl. 460: *Corvus*, Cerinthus.
1870. Butl., Ent. Monthl. Mag. vii. 57: indicates Corvus as the type.

The genus is preoccupied by Chætocnema (Steph., Col. 1831). See Casyapa.

243. CHÆTONEURA.*

1862. Feld., Wien. Ent. Monatschr. vi. 185: *Nearchus* (Hippulus). Sole species, and therefore type.

This name falls before Antigonus (q. v.).

244. CHALYBS.

1816. Hübn., Verz. 76: **Janias**, Telemus, Amyntor (Eryx).
1869. Butl., Cat. Fabr. Lep. 193: uses it for Janias, Telemus, and others.

Janias may be selected as the type.

245. CHAMÆLIMNAS.

1865. Feld., Reise Novara, 304: **Tircis**. Sole species, and therefore type.

Used in same sense by Bates and Kirby.

246. CHARAXES.

1816. Ochs., Schmett. Eur. iii. 18: **Jason** (Jasius). Sole species, and therefore type.

Used in this sense by subsequent authors. See Jasia and Paphia.

247. CHARIDRYAS.

1872. Scudd., Syst. Rev. 26: **Nycteis**, Carlota (Ismeria). Type specified as Nycteis.

248. CHARIS.

1816. Hübn., Verz. 21: Gyas (Gyadis), **Avius** (Ania).
1847. Doubl., List Br. Mus. 16: uses it for a large number of species, including Avius (Anius) of Hübner's list, which therefore becomes the type.
1851. Westw., Gen. Diurn. Lep. 452: uses it for a dozen species, including both of Hübner's.
1867. Bates, Journ. Linn. Soc. Lond. ix. 442: uses it for twenty-nine species, including Avius only of Hübner's.

249. CHILEA.*

1820. Billb., Enum. Ins. 79: proposed in the stead of Libythea, for no reason whatever; of course it falls.

250. CHIONOBAS.*

1832–33 (probably late in 1832). Boisd., Icon. 182: Aello, Norna, Jutta (Jutta, Balder), *Bootes*, Polyxenes (Bore), Œno (Œno, Also).
1833–34 (probably late in 1833). Boisd.-LeC., Lép. Amér. Sept. 214: Jutta (Balder), Bootes, Œno (Œno, Also).
1836. Boisd., Spec. gén., pl. 9 B.: Bootes.

Subsequently used by authors in same sense. But the name must fall before Œneis (q. v.).

251. CHLORIPPE.

1844. [Boisd. in] Doubl., List Br. Mus. 108: Laure (Laura), Laurentia, Zunilda, **Agathina**.
1850. Westw., Gen. Diurn. Lep. 302: gives it as a MS. Boisduvalian synonyme of Apatura.
1870. Boisd., Lép. Guat. 47: claims it as his own, and places in it Laure (Laura).

Boisduval's group consists of two sections, the first two species belonging to one, the last two to the other; the species of the first form the genus Catargyria, and those of the latter may be referred to this name with Agathina for type. See also Doxocopa.

252. CHLORISSES.*

1832–3. Swains., Zoöl. Ill. ii. 89: *Sarpedon*. Sole species, and therefore type.

The name is preoccupied through Chlorissa (Steph., Lep. 1829). See Zetides.

253. CHLOSYNE.

1870. Butl., Cist. Ent. i. 38: proposes this name to take the place of Synchloe Doubl. nec Hübn.

The original species of that group were Erodyle, **Janais**, Tyrinthe (?), and Narva (Bonplandi).

Erodyle, however, was not described until 1864 by Bates (probably using a MS. name of Doubleday in the British Museum). Tyrinthe is still a MS. name, and was omitted from the "Genera," and therefore the type must be either Janais or Narva. Janais as the older name may be taken as the type.

Coatlantona (q. v.) was suggested by Kirby for the same group. It may not be amiss to remark that advance sheets of the portion of Kirby's Catalogue containing this suggestion were received by me in April, 1870, and that Chlosyne was not published until September, 1870. Kirby's Catalogue was not published, however, before the following year, and Butler was previously unaware of the intended change.

254. CHORANTHUS.

1872. Scudd., Syst. Rev. 58: **radians**. Sole species and specified type.

255. CHORIDIS.

1870. Boisd., Lép. Guat. 33: **Peridia**. Sole species, and therefore type.

Will this fall before Aeria?

256. CHORINEA.*

1832. Gray in Griff., An. Kingd., pl. 102, fig. 1: *Licursis* (Xanthippe). Sole species, and therefore type.

But there is another species of the same restricted group which must have been known to Gray, and from which there is scarcely a doubt that he borrowed his generic name to append to his supposed new species; viz., Chorineus. The name should therefore be dropped. It is also very close to Chorinus (Leach, Crust. 1825). See also Zeonia.

257. CHROMA.*

1832. Gray in Griff., An. Kingd., pl. 102, fig. 3: *basilis* (basalis). Sole species, and therefore type.

It cannot be retained, having been preoccupied by Chromis, which is used by Hübner (Lep. 1816), and Cuvier (Fishes, 1817). See also Barbicornis.

258. CHRYSOPHANUS.

1816. Hübn., Verz. 72: Phlæas (Phlæas, Timeus), Helle, Thersamon, Gordius, Hyllus, Alciphron (Hipponoe), **Hippothoe** (Chryseis, Eurybia, Hippothoe), Virgaureæ, Dorilas (Circe).

1841. Westw., Brit. Butt. 91: employs it for Phlæas, Hippothoe (Chryseis, Hippothoe), Dispar, and Virgaureæ.
1850. Steph., Cat. Brit. Lep. 17: the same.
1872. Scudd., Syst. Rev. 35: specifies Hyllus as the type, but the usage of Westwood,' Stephens, and subsequent authors, will not admit of this.

Hippothoe may be taken as the type.

259. CHRYSORYCHIA.

1857. Wallengr. Rhop. Caffr. 44: **Thyra**, Perion (Tjoane).
1858. Ib., K. Vet. Akad. Förh. xv. 80: uses it for Thyra only, which must be taken as the type.

260. CIGARITIS.

1847. Boisd. in Donz., Ann. Soc. Ent. Fr. [2] v. 528: **Zohra**. Sole species, and therefore type.

In speaking of this insect, Donzel says that it belongs to a group of African species, of which Boisduval "*a fait* un genre propre, sous le nom de Cigaritis;" but he does not specify them; nor can I find any mention of the genus by Boisduval himself. Zohra therefore must be considered the type.

1849. Lucas, Expl. Alg. Zoöl. iii. 362: employs it for Siphax, Zohra, and Masinissa, referring the generic name to Boisduval.
1871. Staud., Cat. Lep. Eur. 9: refers the generic name to Lucas.

261. CINCLIDIA.

1816. Hübn., Verz. 29: **Athalia** (Phœbe), Parthenie (Athalia), Dictynna (Orthia).
1850. Steph., Cat. Brit. Lep. 15, 259: uses it for Athalia and Parthenie.
1858. Kirb., List Brit. Rhop.: employs it for the same and another.

Athalia may be taken as type. See also Mellicta and Limnæcia.

262. CIRROCHROA.

1847. Doubl., Gen. Diurn. Lep., pl. 21, fig. 2: **Aoris**. Sole species, and therefore type.
1848. Ib., ib. 157: Aoris and four others, six months later than the plates.

Used in same sense by Felder and Kirby.

263. CISSIA.

1848. Doubl., List Br. Mus. App. 33: **Penelope** (Clarissa) and other species.

All but Penelope, however, bear a query. Penelope must therefore be considered as the type.

It should be noted that there is an allied species (Hesione) called Cissia by Cramer, which was known to Doubleday, having been placed by him in 1844 in Mycalesis. Also that there is a genus Cisia (Boie, Aves, 1826, written Cissa by Gray), which, however, has a distinct derivation.

264. CITHÆRIAS.

1816. Hübn., Verz. 53: Piera (Pieria), Philis (Cissa), **Andromeda,** Nereis.
1865. Herr.-Schaeff., Prodr. i. 55: uses it for Andromeda, Philis, and others not of Hübner's list.
1871. Kirb., Syn. Cat. 36: uses it for several species, including Philis and Andromeda.

Andromeda may be considered as the type. See Callitæra.

265. CLEIS.*

1828–32. Guér., Voy. Coq.: *porticalis.*

It is not a butterfly. See Westw., Gen. Diurn. Lep. ii. 504.

266. CLEODIS.*

1870. Boisd., Lép. Guat. 30.

No species whatever are indicated, nor any type mentioned. It is described and stated to be near Xanthocleis. It is therefore valueless until the author indicates its membership.

267. CLEOSIRIS.*

1836. Boisd., Spec. gén., pl. 7 C.: *Catamita.* Sole species, and therefore type.

This is not a butterfly. See Westw., Gen. Diurn. Lep. ii. 504.

268. CLEROME.

1849? [Boisd. in] Doubl., Gen. Diurn. Lep., pl. 54*: **Arcesilaus.** Sole species, and therefore type.
1851. Boisd. in Westw., Ib. 333: Arcesilaus, Eumeus, Faunula. Boisduval is credited with the name.

It is probable that the plates appeared before the text, but I have no proof of it; in any case, Arcesilaus may be taken as the type. See Faunis.

269. CLOTHILDA.*

1840. Blanch., Hist. Nat. Ins. iii. 440: *Pantherata* (Briaria). Sole species, and therefore type.

1848. Doubl., Gen. Diurn. Lep. 155: uses it for Pantherata and others.

Subsequently used by Felder and Kirby, but the name falls before Anelia (q. v.), Pantherata being strictly congeneric with Numida. See also Synalpe.

270. CLYTIA.*

1832–33. Swains., Zoöl. Ill. ii. 120: Clytia (Clytia, dissimilis), Macareus (Macarius), Panope, specified as types.

As the name of the group is founded upon that of one of the original species included in it, it of course falls. Even if it did not, the name is several times preoccupied, e. g. Hübner (Lep. 1816), Desvoidy (Dipt. 1830), etc.

271. COATLANTONA.

1871. Kirb., Syn. Cat. 178: Saundersii, Paupera, Mediatrix, Lacinia, Melanarge, Janais, misera, Hippodrome, Quehtala, marina, Melitæoides, Erodyle, Pœcile, **Narva**, gaudialis, Perezi, Judith.

Proposed for Synchloe Doubl. nec Hübn.; but Chlosyne had been founded a short time previously for the same purpose. See the remarks under Chlosyne. But all the species of this group cannot be placed in one restricted group, and therefore the name Coatlantona may be retained with Narva for its type.

272. COBALUS.

1816. Hübn., Verz. 115: **Virbius**, Nitocris, Adrastus, triangularis, Phorcus, Hemes, Leucomelas, and a MS. species.

1869. Butl., Cat. Fabr. Lep. 272: uses it for Virbius and other species.

1869. Herr.-Schaeff., Prodr. iii. 77: employs it for nearly eighty species, including Adrastus, triangularis (triangulum), and Phorcus.

1870. Butl., Ent. Monthl. Mag. vii. 92: employs it as a section of Carystus, and specifies Virbius as the type.

273. CŒA.

1816. Hübn., Verz. 48: Varanes (Varanessa), **Acheronta** (Acheronta, Pherecydis).

This has not been used subsequently. Varanes probably belongs to Palla (q.v.); and therefore Acheronta, which is generically distinct from Odius, may be taken as the type.

274. Cœliades.

1816. Hübn., Verz. 106: Forestan, **dubius**, chromus.
    Dubius may be taken as the type.

275. Cœlites.

1851. Boisd. in Westw., Gen. Diurn. Lep. 367: **Nothis**, Epiminthia.
1865. Herr.-Schaeff., Prodr. i. 62: uses it for the same.
1868. Butl., Ent. Monthl. Mag. iv. 195; and Cat. Sat. 111: designates Nothis as type.

276. Cœnonympha.

1816. Hübn., Verz. 65: **Oedipus** (Œdipe), Hero, Dorus (Dorilis), Arcania, Iphis, Corinna (Corynna), Pamphilus (Lylla, Pamphile), Typhon (Philoxena), Leander (Leandra), Philea (Neoclidis).
1843. Herr.-Schaeff., Schmett. Eur. 83: uses it for all of the above.
1844. Doubl., List Br. Mus. 140: makes the same use of it.
1850. Steph., Cat. Brit. Lep. 9, 256: employs it for Typhon (Davus), Pamphilus, Hero, Arcania (Arcanius), and another.
1851. Westw., Gen. Diurn. Lep. 396: uses it for the same and others.
1858. Ramb., Cat. Syst. Lép. Andal. 23: employs it for Pamphilus and Typhon (Davus) only.
1868. Butl., Ent. Monthl. Mag. iv. 194 [Cænonympha]: designates Œdipus (Geticus) as the type.
1871. Kirb., Syn. Cat. 96: employs it for all the Hübnerian species and for others. See Chortobius (p. 293).

277. Cœnophlebia.

1862. Feld., Wien. Ent. Monatschr. vi. 422, note: **Archidona**. Sole species, and therefore type.

278. Cœnyra.*

1865. Hewits., Trans. Ent. Soc. Lond. [3] ii. 281: *Hebe.* Sole species, and therefore type, as subsequently stated by Butler.
1871. Kirb., Syn. Cat. 93: the same.
    This name, however, is preoccupied by the etymologically identical terms Cœnurus (Rud., Worms, 1809), and Cœnura (Big., Dipt. 1857).

279. Cogia.

1870. Butl., Trans. Ent. Soc. Lond. 508: **Hassan**. Sole species, and therefore type.

280. COLÆNIS.

1816. Hübn., Verz. 32: **Julia**, Delila, Lybia, Mercaui.
1848. Doubl., Gen. Diurn. Lep. 148: divides the group into three sections, as follows: I. *a* Delila, Julia, *b* Phærusa; II. Euchroia; III. Dido.
1861. Feld., Neues Lep. 6: divides as follows: I. Phærusa; II. Julia, Delila; III. Dido.
1871. Kirb., Syn. Cat. 147: unites all in one group.
  Julia may be taken as the type.

281. COLIAS.

1807. Fabr., Ill. Mag. vi. 284: I. Palæno, Hyale, Glaucippe; II. rhamni, Cleopatra.
1809. Latr., Gen. Crust. et Ins. iv. 204: uses it for rhamni, Cleopatra, and Hyale.
1810. Ib., Consid. 440: specifies rhamni as the type.
1815. Leach, Edinb. Encycl. 716: restricts the name to Hyale, but erroneously.
1815. Oken, Lehrb. i. 739: makes a similar restriction, and this has been followed by most subsequent authors, whenever they have separated the sections of Fabricius's genus as distinct genera.
1816. Hübn., Verz. 99: employs it for some approximate forms, but including none of Fabricius's.
1820-21. Swains., Zoöl. Ill. i. 5: specifies Eubule (Ebule) as type, erroneously.
1829. Curtis, Brit. Ent., pl. 242: designates Hyale as type, erroneously, as does Westwood in 1840 (Gen. Syn. 87).
1870. Butl., Cist. Ent. i. 43: designates Palæno as type, erroneously.
1872. Ib., ib. i. 66: designates rhamni as type (through Latreille, 1810).
1872. Scudd., Syst. Rev. 38: designates Palæno as type, erroneously.
  See Eurymus, Earina, Gonepteryx, and Gonoptera.

282. COLOBURA.

1820. Billb., Enum. Ins. 79: **Dirce**. Sole species, and therefore type.
  See also Gynæcia.

283. COLOTIS.

1816. Hübn., Verz. 97: Electra, Myrmidone, Croceus (Edusa), Aurora, Chrysotheme, **Amata** (Calais, Cypræa).

1850. Steph., Cat. Brit. Lep. 3, 252: uses it for Croceus (Edusa), Electra, Chrysotheme, and Myrmidone.

But Hyale, a species strictly congeneric with these, had already been taken as the type of Eurymus, and so this action is annulled. Amata must therefore be taken as the type. See also Zerene.

284. COMMA.*

1832. Renn., Consp. 8: *c. album.* Sole species, and therefore type.

Although there is a congeneric species, called comma, it was not named until 1852, and this generic name cannot therefore be affected by it; it falls, however, before Polygonia (q. v.) See also Grapta.

285. COMPSOTERIA.

1870. Hewits., Equat. Lep. iv. 57: **Cascella.** Sole species, and therefore type.
1872. Ib., Exot. Butt. iv.: states that this species belongs to the earlier founded genus Ithomiola, in which case this name falls, and cannot again be employed; but Kirby, in his Synonymical Catalogue, puts them far apart.

286. CONOGNATHUS.*

1862. Feld., Wien. Ent. Monatschr. vi. 181: *Platon.* Sole species, and therefore type.

But the name is preoccupied by Conognatha (Eschsch., Col. 1829).

287. CONSUL.

1806. Hübn., Tent.: **Hippona** (Fabius). Sole species, and therefore type.

See also Fabius, Helicodes, and Protogonius.

288. CORADES.

1848. Boisd. by Doubl. in Hewits., Proc. Zoöl. Soc. Lond. xvi. 115: **Enyo.** Sole species, and therefore type, as subsequently designated by Butler.
1850. Hewits., Ann. Mag. Nat. Hist. [2] vi. 437: uses it for Enyo and others, referring the generic name to Doubleday.

289. CORBULIS.*

1870. Boisd., Lép. Guat. 32: Euphon? (Euphano), Ocalea, Mahela (Neobule), Aletta, Gephira, Nise (Neso, Selene).

The name is preoccupied by Corbula (Brug., Moll. 1791).

290. CORYBANTES.*

1816. Hübn., Verz. 101: Amycus, Dardanus, Icarus, Licus, Syphax, Pylades.

None of these insects are butterflies.

291. CORYBAS.*

1870. Boisd., Lép. Guat. 43: *Tipha* (Typha). Sole species, and therefore type.

This is referred to as a MS. name of Boisduval by Westwood (Gen. Diurn. Lep. 252,—1850), where it is rightly considered as a synonyme of Pyrrhogyra (q. v.), before which it falls, Tipha having become its type in 1844.

292. CORYCIA.*

1822-26. Hübn., Exot. Schmett. ii.: *Appias*. Sole species, and therefore type.

This name is preoccupied by the same name given by Hübner himself (Verz. 1816) to one of the Phalænidæ!

293. CORYDON.*

1869. Hewits., Ill. Diurn. Lep. pt. iv. suppl. 1: *Boisduvalii*. Sole species, and therefore type.

The name is preoccupied in Birds (Less. 1828, Wagl. 1830). See Hewitsonia.

294. COSMOSATYRUS.

1867. Feld., Reise Novara, 495: **leptoneuroides**. Sole species, and therefore type, as specified by Butler.

295. CRASTIA.

1816. Hübn., Verz. 16: **Core**, Climena (Limnoria).

Core may be taken as the type. See Euplœa.

296. CREMNA.

1847. Doubl., List Br. Mus. 14: **Actoris**, and four unpublished species. Actoris must therefore be type.
1851. Westw., Gen. Diurn. Lep. 456: Ceneus, Actoris, and two others; the characters are drawn up from Ceneus.
1867. Bates, Journ. Linn. Soc. Lond. ix. 420: uses it in the same sense.

297. CRENIS.*

1821. Hübn., Index, 2: *Erato* (Brylle). Sole species, and therefore type.
1833. Boisd., Ann. Mus. Hist. 196: madagascariensis.

1847. Ib., Voy. Delegorg. ii. 592 : Drusius (natalensis). These two species have nothing to do with Hübner's genus.

    Doubleday, Wallengren, Butler, and Kirby have used the name in the Boisduvalian sense. The name must fall before Migonitis.

298. Cressida.*

1832–33. Swains., Zoöl. Ill. ii. 94: *Cressida* (Heliconides, Harmonides) designated as type.

    The name being drawn from the species upon which it is founded, it falls. See Eurycus.

299. Cricosoma.

1865. Feld., Reise Novara, 292: **leopardinum**. Sole species, and therefore type.

    Used in same sense by Bates and Kirby. Although the name is very close to Cricostoma (Klein, Moll. 1753), it differs etymologically.

300. Crocozona.

1865. Feld., Reise Novara, 296 : **Pheretima**. Sole species, and therefore type.

301. Cupha.

1820. Billb., Enum. Ins. 79 : **Erymanthis**. Sole species, and therefore type. See Messaras.

302. Cupido.

1801. Schrank, Fauna Boica, ii. i. 153, 206 : I. Virgaureæ, Hippothoe (Hippothoe, Chryseis), Phlæas, Dorilas (Circe) ; II. **Arion**, Alcon, Semiargus (Acis), Damon, Cyllarus (Damœtas), Argiolus, Chiron (Eumedon), Corydon, Thetis (Adonis), Alexis, Corydon (Agestis), Argus, Battus, Argiades (Puer), Alsus (Puer); III. rubi, betulæ, quercus, pruni, spini.

1816. Hübn., Verz. 77: uses it for Hymen (Liger), Amor, and Chrysus. These have no generic connection with any of Schrank's species, but the last of them is closely allied to the species Cupido Linn., which may have been the cause of Hübner's selection.

1871. Kirb., Syn. Cat. 345 : uses it in place of Lycæna of most modern authors, including some three hundred species, and among them all of Schrank's second section.

    The name may be retained for the group represented by the first two species of the second section, with Arion for the type. [See p. 293.]

303. CURETIS.

1816. Hübn., Verz. 102: **Thetys** (Æsopus), Pasiphae (Ormenus).
1871. Kirb., Syn. Cat. 418: uses it for Thetys and Bulis.
    Thetys therefore becomes the type. See also Anops, Candalides, and Phædra.

304. CYANE.*

1861. Feld., Neues Lep. 22: *Leprieurii*. Sole species, and therefore type.
    The name is, however, preoccupied by Cyanea (Pér. et Les., Acal. 1809).

305. CYANIRIS.

1816. Dalm., Vetensk. Acad. Handl. xxxvii. 63, 94: Arion, Alcon, Cyllarus, Semiargus (Argianus), **Argiolus**, Alsus, Icarius, Thetis (Adonis), Icarus (Alexis), Alexis (Agestis), Chiron (Eumedon), Optilete, Battus, Argus.
1820. Billb., Enum. Ins. 80: uses it for all of Dalman's excepting Alcon, and for several additional species.
1872. Scudd., Syst. Rev. 34: indicates Argiolus as type. [See p. 293.]

306. CYBDELIS.

1836. Boisd., Spec. gén., pl. 5 B.: **Phæsyle** (Phæsila). Sole species, and therefore type.
1844. Doubl., List Br. Mus. 89: uses it for Sydonia and others, but without including Phæsyle.
1849. Ib., Gen. Diurn. Lep. 217: uses it for three species, including Phaesyle (Phæsila).

307. CYCLOGRAMMA.

1847. Doubl., Gen. Diurn. Lep., pl. 27: **Pandama**. Sole species, and therefore type.
1848. Ib., ib. 219: the same, and a MS. species.

308. CYCLOPIDES.

1816. Hübn., Verz. 111: Morpheus (Steropes), Palæmon (Brontes), Sylvius, **Metis**, Coras.
1830. Steph., Cat. Brit. Lep. 22, 262: uses it for Palæmon (Paniscus) and Sylvius.
1861. Staud., Cat. 15: employs it for Morpheus (Steropes).
1866. Trim., Rhop. Afr. Austr. 292: employs it for Metis and other African species.

1870. Butl., Ent. Monthl. Mag. vii. 96: indicates Morpheus (Steropes) as type, but erroneously, because Morpheus was taken by Dumeril as the type of Heteropterus in 1823. By a similar error,

1872. Scudd., Syst. Rev. 54: indicates Morpheus (Steropes) as the type.

Palæmon or Sylvius cannot be taken as the type, as would follow from Stephens's action in 1850, because Palæmon must be taken for Pamphila (q. v.): Metis may therefore be selected. See also Erynnis.

309. CYCNUS.

1816. Hübn., Verz. 81: Chiton, **Phaleros** (Agis, Phaleros), Linus (Ætolus).

Phaleros may be taken as the type.

310. CYLLO.

1832. Boisd., Voy. Astrol. 140: amabilis, Constantia, **Leda**.
1844. Doubl., List Br. Mus. 120: employs it for the two latter species and others.
1851. Westw., Gen. Diurn. Lep. 360: designates Leda (Leda, Banksia) as type.

311. CYLLOGENES.

1868. Butl., Ent. Monthl. Mag. iv. 194: **Suradeva**. Sole species and designated type.

312. CYLLONIUM.* (Fossil.)

1854. Westw., Quart. Journ. Geol. Soc. Lond. 395–6: Boisduvalianum, Hewitsonianum.

The latter is not a butterfly, and it is exceedingly doubtful if the former can be so considered. The genus is uncharacterized, but the species are figured; they are, however, so fragmentary that it would be impossible to trace any generic characters from them.

313. CYLLOPSIS.

1869. Feld., Verh. zoöl.-bot. Gesellsch. Wien. xix. 474: **Hedemanni**. Sole species, and therefore type.

314. CYMÆNES.

1872. Scudd., Syst. Rev. 61: **tripuncta**, malitiosa. Tripuncta specified as type.

315. CYMATOGRAMMA.

1849. Doubl., Gen. Diurn. Lep. pl. 49: **Echerus**. Sole species, and therefore type.
1850. Ib., in Westw. Gen. Diurn. Lep. 315: the same.

316. CYMOTHOE.*

1816. Hübn., Verz. 39.: Cænis (Amphicede), Althea, Aconthea.
1871. Kirb., Syn. Cat. 251: employed for Cænis, Althea, and many others.

Preoccupied by Cymothoa (Fabr., Crust. 1798).

317. CYNTHIA.

1807. Fabr., Ill. Mag. vi. 281: I. **Arsinoe**, interrogationis; II. Œnone, Jatrophæ, cardui, Statilinus (Allionia).
1815. Oken, Lehrb. i. 737: employs it erroneously for Maturna, Cynthia, etc.
1827. Steph., Ill. Brit. Ent. Haust. 47: restricts it to cardui and Vellida (hamptsteadiensis).
1828. Horsf., Descr. Cat. Lep. E. Ind. Co., expl. pl.: cardui only.
1840. Westw., Gen. Syn. 87: specifies cardui as type.
1841. Westw., Brit. Butt. 56: uses it for cardui, Huntera, and Vellida (hamptsteadiensis).
1849. Doubl., Gen. Diurn. Lep. 212: restricts it to Arsinoe only.
1871. Kirb., Syn. Cat. 152: restricts it to Arsinoe and Erota.
1872. Crotch, Cist. Ent. i. 66: says that cardui is type, through Horsfield's action in 1828.

Cardui would be type, but that it is strictly congeneric with Atalanta which was previously (Latr. 1810) designated as type of Vanessa. Arsinoe must therefore be taken as the type.

318. CYRENIA.

1851. Westw., Gen. Diurn. Lep. 434: **Martia**. Sole species, and therefore type.

Should this name be dropped as preoccupied? Cyrene has been used in Fishes (Heck. 1840), and Hemiptera (Westw.! 1841).

319. CYRESTIS.

1832. Boisd., Voy. Astrol. 117: **Thyonneus** (Thyoneus), Acilia.
1833. Ib., Ann. Mus. Hist. Nat. 190: elegans.

It has since been used in the same sense by different authors, as Doubleday, Westwood, etc.

1861. Feld., Neues Lep. 24: divides the group into two sections, to the first of which he refers Thyonneus, and to the second Risa and Rahria.

Thyonneus may then be taken as the type.

320. Cystineura.*

1836. Boisd., Spec. gén., pl. 5 B.: *Dorcas* (Hersilia). Sole species, and therefore type.

<small>Similarly used by subsequent authors. The name must fall before Mestra (q. v.).</small>

321. Dædalma.

1858. Hewits., Exot. Butt. ii. 85: **Dinias,** Drusilla, Doræte, Drymæa.
1867. Butler, Ann. Mag. Nat. Hist. [3] xx. 268; Cat. Sat. 183: specifies Dinias as type.

322. Damis.*

1832. Boisd., Voy. Astrol. 67: Cyanea (Epicoritus), Euchylas (Coritus), Danis (Scbœ).

<small>This name falls because founded on a specific name, Damis being one of the synonymes of Danis. See Danis and Thysonotis.</small>

323. Damora.

1851. Nordm., Bull. Mosc. xxiv. ii. 439: **Sagana** (Paulina). Sole species, and therefore type.
1861. Feld., Neues Lep. 10: uses the name as a division of Argynnis.

324. Danaida.

1805. Latr., Sonn. Buff. xiv. 108: **Plexippus.** Sole species, and therefore type.
1830? Guér., Icon. Règne An. 474, pl. 77: Kadu (Eunica) and another.

<small>This name is preoccupied only in botany, and ought to be restored. See Danaus.</small>

325. Danaus.*

1809. Latr., Gen. Crust. et Ins. iv. 201: I. *Plexippus,* similis, Midamus; II. Idea.
1810. Ib., Consid. 440: specifies Idea and Plexippus as types.
1815. Oken, Lehrb. i. 723: uses it for cratægi and allies, the Danai candidi of Linné.
1819. God., Encycl. méth. ix. 172: uses it with the spelling Danais (which has clung to it through the writings of most subsequent authors) for Latreille's first group. Consequently Plexippus is the type.

1872. Crotch, Cist. Ent. i. 60: refers it, as Oken did, to Linné and says that the type was fixed by Cuvier in 1799 (Tabl. Élém.) as brassicæ.

> Linné, however, used no such word in a generic sense, but divided his genus Papilio into sections, to which he gave plural names, Danai, etc.; these again into subsections, such as Danai festivi, etc.; we must therefore disregard them altogether in treating of genera. Cuvier used it in the same plural form (Danai), but referred rapæ as well as brassicæ to it.

1872. Scudd., Syst. Rev. 7: specifies Plexippus as type.

> Inasmuch as the name Danaus was proposed by Latreille to supplant his own Danaida (because preoccupied in botany, and of which Plexippus was the type), and since he subsequently specified Plexippus as one of the types in 1810, before any action had affected the Fabrician Euplœa, Plexippus would have to be considered the type of this genus, could it stand. See Danaida.

326. DANIS.*

1807. Fabr., Ill. Mag. vi. 286: no species whatever indicated.
1815. Oken, Lehrb. i. 722: gives it as a division of Emesis, with no mention of species beyond the remark "vier Arten," which is copied from Fabricius.
1820. Billb., Enum. Ins. 80: changes the name for no apparent reason to Hadothera (q. v.).
1852. Westw., Gen. Diurn. Lep. 497: uses it for several species, commencing with Danis (Sebæ).

> Very probably Westwood was right, as Fabricius in several instances used specific names for genera containing them. But we cannot possibly determine this fact. If it were true, the name would fall from its illegal derivation. If it were not, it would have to be dropped for want of support, or possible fixity. See Damis and Thysonotis.

327. DAPTONOURA.*

1869. Butl., Cat. Fabr. Lep. 209: Lycimnia, Salacia. Lycimnia specified as type.

> This name falls before Melete (q. v.).

328. DARPA.

1865. Moore, Proc. Zoöl. Soc. Lond. 781: **Hanria**. Sole species, and therefore type.

329. DASYOMMA.*

1860. Feld., Wien. Ent. Monatschr. iv. 401: *fuscum*. Sole species, and therefore type.

> This name is preoccupied in Diptera (Mng. 1840).

330. DASYOPHTHALMA.

1851. Westw., Gen. Diurn. Lep. 343: **Rusina**, Creusa.

Used since in same sense by Herrich-Schaeffer, Hewitson, and Kirby. Rusina may be taken as the type.

331. DEBIS.

1849. [Boisd. in] Doubl., Gen. Diurn. Lep. pl. 61: **Samio**. Sole species, and therefore type.

1851. Boisd. in Westw., Gen. Diurn. Lep. 358: uses it for Europa and eight others, including Samio.

Westwood designates Europa as type, but erroneously; for the plates had then been published two years, and there is no indication that Boisduval, whose statement alone would have had force, intended Europa for the type. No Boisduvalian species are placed in the genus by Westwood, and moreover Europa is the type of Lethe, so that Samio must certainly be taken as the type. If, however, Samio is generically identical with Europa, then Debis falls before Lethe (q. v.).

332. DELIAS.

1816. Hübn., Verz. 91: **Egialea** (Tyche, Apriate), Pasithoe (Pasithoe, Porsenna).

1870. Butl., Cist. Ent. i. 34: specifies Egialea as the type.
See Thyca.

333. DELONEURA.

1868. Trim., Trans. Ent. Soc. Lond. 81: **immaculata**. Sole species, and therefore type.

1871. Kirb., Syn. Cat. 426: uses it in the same sense.

334. DERCAS.

1847. Boisd. in Doubl., Gen. Diurn. Lep. 70: **Verhuellii**. Sole species, and therefore type, as subsequently indicated by Butler.

335. DESMOZONA.*

1836. Boisd., Spec. gén., pl. 2 B., 5 C.: Mantus (Manthus), Acherois.
1868. Bates, Journ. Linn. Soc. Lond. ix. 451: uses it for thirty-five species, including both of Boisduval's.

The name falls before Peplia. See also Heliochlæna and Nymphidium.

336. DEUDORIX.

1863. Hewits., Ill. Diurn. Lep. i. 16: Eryx (Amyntor), Perse, Epirus (Epirus, Despœna), Eos, Lexias, Domitia, **Epijarbas**, Diovis, Xenophon, Diœtas, Pheretima, Petosiris, Melampus, Phranga, Sphinx (Varuna), Elcia, Manea, Nissa, Orseis, Nasaka, Chozeba, Isocrates, Antu, Galathea, Timoleon, Mæcenas (Timoleon). Epijarbas is designated as the type.

Subsequently used by Herrich-Schaeffer, Butler, and Kirby. See Artipe.

337. DIADEMA.*

1832. Boisd., Voy. Astrol. i. 135: *Bolina* (Lasinassa), Alimena, Pandarus (Pipleis).
1833. Ib., Ann. Mus. Hist. Nat. 187: Bolina, dubius (dubia).

Bolina therefore becomes the type, and in this sense the genus is used by Doubleday, Westwood, and Felder; but the name is preoccupied in Crustacea (Schum. 1817) and Echinoderms (Gray, 1825). See Esoptria and Apatura.

338. DIÆTHRIA.

1820. Billb., Enum. Ins. 78: **Clymena.** Sole species, and therefore type. See Callicore.

339. DICALLANEURA.

1867. Butl., Proc. Zoöl. Soc. Lond. 37: **pulchra**, decorata.
Pulchra may be taken as the type.

340. DICHORRAGIA.

1868. Butl., Proc. Zoöl. Soc. Lond. 614: **Nesimachus.** Sole species and designated type.

341. DIDONIS.

1816. Hübn., Verz. 17: **Vitellia**, Biblis.
1844. Doubl., List Br. Mus. 144: uses it for Biblis and its allies, in which sense it has since been used by Westwood, Herrich-Schaeffer, and Kirby.

But Billberg had earlier (1820) taken Biblis as the type of Zonaga, and hence Vitellia must be used as the type.

342. DILIPA.

'1858. Moore, Cat. Lep. East Ind. Co. i. 201: **Morgiana.** Sole species, and therefore type.

343. DIONE.

1816. Hübn., Verz. 31 : vanillæ, **Juno**.
Juno may be taken as the type.

344. DIOPHTHALMA.

1836. Boisd., Spec. gén., pl. 2 B., 5 C.: Sifia, **Telegone**.
Telegone may be taken as type.

345. DIORINA.

1837. Boisd. in Mor., Ann. Soc. Ent. Fr. vi. 421: **Periander** (Laonome). Sole species, and therefore type.

Since used in same sense; but frequently (as by Doubleday, Bates, and Kirby), with the incorrect spelling Diorhina. See also Rodinia and Rhetus.

346. DIPSAS.*

1847. Doubl., List Br. Mus. 25: Ataxus, *Syla* (Pholus), (both inedited).
1852. Westw., Gen. Diurn. Lep. 479: Syla (Sila), Ataxus, and ten others. Syla specified as type.
1865. Hewits., Ill. Diurn. Lep. 64: also specifies Syla and Ataxus as types.

The genus is however preoccupied in Reptiles (Lam. 1768) and Mollusks (Leach, 1814).

347. DIRA.

1816. Hübn., Verz. 60: Roxelana (Roxelane), Megæra, Mæra, **Clytus** (Clyte).
1850. Steph., Cat. Brit. Lep. 6: uses it for Megæra.

Clytus must be taken as the type, as the other species fall into Lasiommata and Pararge. See also Amecera, Leptoneura, and Maniola.

348. DIRCENNA.

1847 (Aug.). Doubl., Gen. Diurn. Lep. pl. 17: **Jemima** (Iambe).
Sole species, and therefore type.
1847 (Nov.). Ib., ib. 119: Jemima (Iambe), and other MS. or queried names.
1862. Bates, Linn. Trans. xxiii. 520: employs it for eight species, including the above.
1871. Kirb., Syn. Cat. 20: follows Bates.

349. DISCOPHORA.

1836. Boisd., Spec. gén., pl. 4 A., 8 B.: **Celinde** (Menetho), Sondiaca.

In the explanation of the plates, the name is spelled as above, but on Pl. 4 A. it is spelled Discophorus. Subsequently used in same sense by Doubleday, Westwood, and Kirby. Celinde may be taken as the type. The name is very close to Discopora (Lam., Pol. 1816).

350. DISMORPHIA.

1816. Hübn., Verz. 10: **Laia** (Laja), Amphione.
1870. Butl., Cist. Ent. 39, 54: Laia (Laja) specified as the type.

351. DODONA.

1861. Hewits., Exot. Butt. ii. 91: **Durga**, Egeon.

Used in same sense by Herrich-Schaeffer and Kirby. Durga may be taken as the type.

352. DOLESCHALLIA.

1861. Feld., Neues Lep. 14: **Bisaltide** (Polibete, Bisaltide). Sole species, and therefore type.

353. DORILA.*

1832. Gray, Griff., An. Kingd., pl. 102, fig. 2: *Asteris*. Sole species, and therefore type.

Preoccupied by Dorylus (Latr., Hym. 1802), and doubtless also it owes its origin to the specific name of its close ally, Dorilas. See Syrmatia.

354. DORITIS.

1807. Fabr., Ill. Mag. vi. 283: Apollo, **Mnemosyne.**
1815. Leach, Edinb. Encycl. 716: uses it for Apollo only; but Apollo was selected in 1810 as the type of the earlier genus Parnassius (q. v.), so that Mnemosyne, which is generically distinct from Apollo, must be taken as type.
1816. Dalm., Vetensk. Acad. Handl. xxxvii. 60: specifies Apollo as type.
1816. Hübn., Verz. 89: employs it for Apollinus (Thia) only, but incorrectly.
1832. Dup., Pap. France, Diurn. Suppl. 380: also restricts it to Apollinus (Apollina), in which he is followed wrongly by Boisduval and others; Boisduval even says of the genus, "Établi par nous dans notre Icones"!
1840. Westw., Gen. Syn. 87: specifies Apollo as type.
1872. Crotch, Cist. Ent. i. 66: says that Apollina is type, through Hübner, 1816.

See Archon, Parnassius, and Therius.

355. Doxocopa.

1816. Hübn., Verz. 49: Erminea, Iris, Ilia (Ilia, Astasia), Agathina, **Polyxena** (Epilais).
1865. Herr.-Schaeff., Prodr. i. 80: uses it for Idyia, Celtis, Argus, none of them Hübner's species, although congeneric with Agathina.
1872. Scudd., Syst. Rev. 9: in accordance with Herrich-Schaeffer's use of the name, specifies Agathina as type.

But it cannot so be considered, since Agathina must be referred to Chlorippe (q. v.). Erminea has been taken as the type of Apaturina, Iris and Ilia belong to Potamis, and hence this group must be restricted to Polyxena and its allies.

356. Drucina.

1872. Butl., Cist. Ent. i. 72: **Leonata** specified as type.

357. Drusilla.*

1820-21. Swains., Zoöl. Ill. i. i. 11: *Urania* (Jairus), Horsfieldii. Urania (Jairus) specified as type.

Used in same sense by Westwood: but the name falls before Tænaris, and is preoccupied in Coleoptera (Leach, 1819).

358. Dryas.

1806. Hübn., Tent. 1: **Paphia.** Sole species, and therefore type.
1847. Boisd., Voy. Deleg. ii. 588: uses it for Leda, which has nothing to do with the Hübnerian genus, but belongs to a distinct family.
1865. Feld., Reise Novara, 305: uses it for Cinaron, which has nothing to do with either the Hübnerian or the Boisduvalian genus! See Aculhua.

Wallengren has followed Boisduval; Bates has followed Felder; and Herrich-Schaeffer, in his Prodromus, has followed both the one and the other! See Argynnis.

359. D'Urbania.

1862. Trim., Trans. Ent. Soc. Lond. [3] i. 400: **Amakosa.** Sole species, and therefore type.

360. Dyctis.

1832. Boisd., Voy. Astrol. 138: **Agondas.** Sole species, and therefore type.
1851. Westw., Gen. Diurn. Lep. 353: the same.

361. DYNAMINE.*

1816. Hübn., Verz. 41 : *Mylitta* (Postuerta, Mylitta). Sole species, and therefore type.

> Subsequently used by Kirby, but preoccupied by Dynamena (Lam., Pol. 1812).

362. DYNASTOR.

1849. Doubl., Gen. Diurn. Lep., pl. 58 : **Napoleon.** Sole species, and therefore type.

1851. Doubl. in Westw., Gen. Diurn. Lep. 346 : Napoleon (Napoleo), Darius.

> The plates being in advance of the text, Napoleon must be the type. See Megastes.

363. DYNOTHEA.

1866. Reak., Proc. Ent. Soc. Phil. v. 222 : **Lycaste,** Salapia, Terra, Avella, Diasia.

> Lycaste may be taken as the type.

364. DYSENICS.*

1872. Scudd., Syst. Rev. 46 : *albicilla,* cruentus. Albicilla specified as the type.

> As albicilla is congeneric with Palemon, the necessary type of Phocides, this name must fall.

365. DYSMATHIA.

1867. Bates, Journ. Linn. Soc. Lond. ix. 382 : **Portia,** costalis.

> Portia may be taken as type.

366. EAGRIS.

1863. Guén. in Maill., Reun. ii. Lep. 19 : **Sabadius.** Sole species, and therefore type.

367. EANTIS.

1836. Boisd., Spec. gén., pl. 9 B. : **Thraso.** Sole species, and therefore type.

368. EARINA.*

1839. Speyer, Isis, 98 : rhamni, Cleopatra.

> This name is preoccupied by Earinus (Wesm., Hym. 1837). See also Colias, Gonepteryx, and Gonoptera.

369. Ecaudati.*

1860. Koch, Stett. Ent. Zeit. xxi. 230: Memnon and a host of others.

The objections urged against Caudati (q. v.) of the same author would consign this also to oblivion, even if the name were not preoccupied in Reptiles (Dum. 1806), as well as in connection with the swallow-tails (Billberg, 1820). It is also used, but not in a generic sense, by Swainson (Zoöl. Ill.) for a division of swallow-tails.

370. Echenais.

1816. Hübn., Verz. 19: Arius (Aria), Misenes (Misenessa), **Aristus** (Leucophæa), Emylius (Emylia), pseudocrispus (Luciane, Crispa), Thelephus (Alphæa), Chia.

1871. Kirb., Syn. Cat. 325: employs this for fifteen species, including, of Hübner's, only Aristus, which thereby becomes the type.

The name is very close to Echeneis (Linn., Fishes, 1748).

371. Ectima.

1848 (Nov.) Doubl., Gen. Diurn. Lep., pl. 42: **Iona.** Sole species, and therefore type.

1849 (July). Ib., ib. 227: Liria.

The text states that a species distinct from Liria "will be figured" in one of the supplementary plates. It will be noted that both text (which is Doubleday's) and plate were published after Doubleday's death. Although the description was drawn up from Liria, Iona must be taken as the type.

372. Elina.

1852. Blanch. in Gay's Chili, vii. 28: **Vanessoides**, Lefebvrei (Montrolii).

1868. Butl., Ent. Monthl. Mag. iv. 194; and Cat. Sat. 64: designates Vanessoides as type.

373. Elodina.

1865. Feld., Reise Novara, 215: **Egnatia** (Therasia), Hypatia.
1867. Herr.-Schaeff., Prodr. ii. 8: uses it in the same sense.
. 1870. Butl., Cist. Ent. i. 34, 40: designates *Egnasia* (Therasia) as the type.

374. Elymnias.

1816. Hübn., Verz. 37: undularis (Protogenia, Jynx, undularis), **Lais**, Enotrea (Enothrea), Ariadne.

- 1869. Butl., Cat. Fabr. Lep. 38: employs it for Vitellia, undularis, Panthera, and Lais.
- 1871. Kirb., Syn. Cat. 112: uses it in the same sense.

Lais may be taken as the type.

375. EMESIS.

- 1807. Fabr., Ill. Mag. vi. 287: **Fatima** (Ovidius), Absolon (Absalon).
- 1815. Oken, Lehrb. i. 722: enlarges it greatly, making it include many other of the Fabrician genera, but in Emesis proper places the Fabrician species and others.
- 1816. Hübn., Verz. 21: refers four species to it allied to Fatima, but places that in his Polystichtis. See Mesene.
- 1820. Billb., Enum. Ins. 81: changes the name, for no apparent reason, to Tapina.
- 1818. Hoffm., Wied. Zoöl. Mag. i. ii. 98: employs it for Fatima (Ovidius), and many others, not including Absolon, so that Fatima becomes the type.
- 1840. Blanch.-Brullé, Hist. Nat. Ins. iii. 466: specify Crœsus as type, of course erroneously.
- 1847. Doubl., List Br. Mus. 9: employs it for Lucinda, Mandana, (Arminius), and Fatima, besides some MS. species.
- 1851. Westw., Gen. Diurn. Lep. 421, 446: specifies Fatima (Ovidius) as type.
- 1867. Bates, Journ. Linn. Soc. Lond. ix. 436: employs it for Lucinda and others, including neither of the species of Fabricius.
- 1871. Kirb., Syn. Cat. 312: follows Bates in general, but includes Fatima.
- 1872. Crotch, Cist. Ent. i. 66: says Fatima (Ovidius) is type through Westwood in 1850. See Nelone.

376. ENANTIA.

- 1816. Hübn., Verz. 96: Melite, **Licinia** (Lininia).

Licinia may be taken as the type. See Licinia.

377. ENDOPOGON.

- 1864. Boisd. in Feld., Spec. Lep. 4: **Sesostris** (Sesostris, Zestos, Tarquinius), Childrenæ, Vertumnus (Vertumnus, Cutora, Iphidamas, Erithalion, Zeuxis, Alyattes, Rhamases), Anchises, Telmosis, Erlaces, Phosphorus, Cyphotes (Hierocles, Eteocles, Lycomes), Nephalion, Idalion,

Polyzelus, Arcas (Xenares, Arcas), Toxaris (Toxaris, Anacharsis), Cymochles (Cymochles, Orbignyanus), Serapis (Serapis, Osyris), Pomares.

Sesostris may be taken as the type.

378. ENDYMION.*

1832–33. Swains, Zoöl. Ill. ii. 85: *regalis*. Sole species, and therefore type.

But Endymion is one of the synonymes of regalis, so that this name must be dropped. See Eucharia, Evenus, and Arcas.

379. ENISPE.

1848. Doubl., Gen. Diurn. Lep., pl. 40: **Enthymius**. Sole species, and therefore type.

1850. Ib. in Westw., Gen. Diurn. Lep. 292: the same.

380. ENODIA.

1816. Hübn., Verz. 61; **Portlandia** (Andromacha), Dejanira, Hyperanthus (Hyperanthe).

1844. Doubl., List Br. Mus. 136: uses it for Alope and Hyperanthus.

In this sense the genus is used by Westwood, Stephens, and Kirby (List), but Hyperanthus was virtually the type of Hipparchia in 1820.

1872. Scudd., Syst. Rev. 5: designates Portlandia as the type.

381. ENOPE.*

1858. Moore, Cat. Lep. E. Ind. Co. i. 228: Pulaha, Bhadra.

According to Kirby (in litt.) this name is preoccupied in Lepidoptera (Walk. 1854). See Neope.

382. ENTHEUS.

1816. Hübn., Verz. 114: **Peleus**. Sole species, and therefore type.

1871. Kirb., Syn. Cat. 579: uses it for more than twenty species, including Peleus. See Phareas and Peleus.

383. EPARGYREUS.

1816. Hübn., Verz. 105: Prodicus, **Tityrus** (Clarus), Pomus (Comus), Evadnes, Epitus, Brino, Orchamus.

1869. Butl., Cat. Fabr. Lep. 275: uses it for Mathias and others, none of which are mentioned by Hübner.

1872. Scudd., Syst. Rev. 49: specifies Tityrus as type.

384. EPHYRIADES.

1816. Hübn., Verz. 111: **Otreus**, Folus, Tryxus, Asychis.
   Otreus may be taken as the type.

385. EPICALIA.*

1844. [Boisd. in] Doubl., List Br. Mus. 90: Acontius (Antiochus), Numilia (Numilius), Obrinus (Ancæa).
1850. Boisd. in Westw., Gen. Diurn. Lep. 256: uses it in same sense.
1861. Feld., Neues Lep. 17: I. Acontius; II. species not mentioned by Doubleday; III. Obrinus (Ancæa), and another.
1870. Boisd., Lép. Guat. 40: claims it as his own, and refers to it Nyctimus, Antinoe, and Numilia.

   The name is preoccupied through Epicallia, used in Lepidoptera (Hübn. 1816), and Epicalla, used in Coleoptera (Dej. 1833). All have the same derivation.

386. EPIGEA.*

1816. Hübn., Verz. 62: Euryale (Adyte, Euryale), Ligea, Embla, Medea, Pyrrha.
1850. Steph., Cat. Brit. Lep. 8: uses it for Ligea and Medea (Blandina).

   But the name falls before Erebia. See also Gorgo, Marica, Syngea, Phorcis, and Oreina.

387. EPINEPHELE.*

1816. Hübn., Verz. 59: *Jurtina* (Janira), Lycaon (Eudora), Clymene (Synclimene).
1843. Herr.-Schaeff., Schmett. Eur. i. 81: uses it for the first two of these and for others.
1850. Steph., Cat. Brit. Lep. 7 [Epinephila]: uses it for Jurtina (Janira).
1858. Kirb., List Brit. Rhop.: the same.
1868. Butl., Ent. Monthl. Mag. iv. 194; and Cat. Sat. 64: designates Jurtina (Janira) as type.

   The name, however, is preoccupied by Epinephelus (Bloch-Schneid., Fishes, 1801).

388. EPINETES.*

1820. Billb., Enum. Ins. 77: Ceres (Sebethis MS.), Isabella, Calliope, Polymnia, Psidii, diaphanus, and some MS. species.
   A heterogeneous group which would best be left unused.

389. Epiphile.

1844. [Boisd. in] Doubl., List Br. Mus. 90 : **Orea**, Laothoe (Merione, Liberia).
1849. Boisd. in Doubl., Gen. Diurn. Lep. 224 : Orea, Chrysites, Lampethusa,? Laothoe.
1870. Boisd., Lép. Guat. 40 : Ariadne, Chrysites, Epicaste, Adrasta.
1871. Kirb., Syn. Cat. 201 : employs it for all the above, excepting Laothoe and Ariadne, and for others.

Orea may therefore be taken as the type.

390. Epitola.

1852. Boisd. in Westw., Gen. Diurn. Lep. 470 : **Elion**. Sole species, and therefore type.

391. Erebia.

1816. Dalm., Vetensk. Acad. Handl. xxxvii. 58, 79 : I. Ægeria (Egeria), Mæra, Megæra; II. Semele, Agave (Hippolyte), Norna, Polyxenes (Bore), Hyperanthus, Medusa, **Ligea**, Manto, Embla, Jurtina (Janira), Typhon (Davus), Pamphilus, Hero, Arcania, Iphis. Ligea is specified as type.
1832-3. Boisd., Icones, 147 : uses it for Pronoe (Neorides) and others, including a mention of Ligea.
1843. Herr.-Schaeff., Schmett. Eur. 55 : refers to it a large number of species, including Ligea.
1844. Doubl., List Br. Mus. 123 : employs it for a large number, including Ligea.
1850. Westw., Gen. Diurn. Lep. 376 : makes similar use of it.
1868. Butl., Ent. Monthl. Mag. iv. 194; and Cat. Sat. 72 : specifies Ligea as type.

See Gorgo, Marica, Syngea, Phorcis, Epigea, Oreina, and Maniola.

392. Eresia.

1836. Boisd., Spec. gén., pl. 7 B.: **Eunice** (Eunica). Sole species, and therefore type.
1844. Doubl., List Br. Mus. 64: uses it in this sense.
1848. Ib., Gen. Diurn. Lep. 182: the same extended.
1861. Feld., Neues Lep. 10 and App.: refers to it Nycteis, Ismeria, and others.

The name is very near to Eresus (Walck. Arachn. 1805).

393. ERGOLIS.

1836. Boisd., Spec. gén. pl. 4 A: **Ariadne** (larva and pupa only figured). Sole species, and therefore type.

Used in same sense by Doubleday, Westwood, and Kirby. See Ariadne.

394. ERIBŒA.

1816. Hübn., Verz. 46: Brutus (Bruta), Jason (Unedonis), Pelias (Pelopia), Lucretia, Castor (Castoris), Pollux (Pollussa), Æclus (Aile), Tiridates (Tiridatis), Athamas (Athamis), **Etheocles** (Etheoclessa), Xiphares (Thyestessa, Xypharis), Pyrrhus (Pyrrichia), Euryalus (Euriale).

Is this name too near Erebia (Dalm., Lep. 1816) to be used? If not, Etheocles may be considered the type.

395. ERINA.*

1832–33. Swains., Zoöl. Ill. ii. 134: Xanthospilos (pulchella), Erinus, ignita. Erinus specified as typical.

The generic name, being based upon it, must drop. See Holochila and Polycyma.

396. ERITES.

1851. Boisd. in Westw., Gen. Diurn. Lep. 392: **Medura** (Madura). Sole species, and therefore type, as stated by Butler.

397. EROESSA.

1847. Doubl., Gen. Diurn. Lep. 56: **chilensis**. Sole species, and therefore type, as stated by Butler.

398. ERONIA.

1822–26. Hübn., Exot. Schmett. ii.: **Cleodora**. Sole species, and therefore type, as specified by Butler.

1836. Boisd., Spec. gén. 604: the same.

The genus has been used in the same sense by authors.

399. ERORA.

1872. Scudd., Syst. Rev. 32: **læta**. Sole species and designated type.

400. EROTION.*

1820. Dalm. in Billb., Enum. Ins. 80. *Cupido*. Sole species, and therefore type.

The name falls before Helicopis. See also Hexuopteris.

401. ERYCIDES.

1816. Hübn., Verz. 110: **Pygmalion** (Pigmalion), Gnetus (Megalesius).

1852. Westw., Gen. Diurn. Lep. 509: uses it for seven species, including Pygmalion.
1869. Butl., Cat. Fabr. Lep. 266: employs it for Palemon only, not specified by Hübner.
1869. Herr.-Schaeff., Prodr. iii. 59: refers a great many species to it, including Pygmalion and Palemon.
1870. Butl., Ent. Monthl. Mag. vii. 92: employs it not only for Palemon, but for many others, including Pigmalion.
1871. Kirb., Syn. Cat. 587: places Pygmalion in it, with nearly thirty others.
1872. Scudd., Syst. Rev. 46: specifies Pygmalion (Pigmalion) as the type.

402. ERYCINA.*

1807. Fabr., Ill. Mag. vi. 286: *Melibœus*, Lysippus, Orsilochus.
1809. Latr., Gen. Crust. et Ins. iv. 205: extends the group, including in it all of the above.
1810. Ib., Consid. 440: Lamis, Fatima, Melander, Lysippus, and Melibœus are specified as types. One of the last two of these must therefore be chosen.
1815. Oken, Lehrb. i. 722: gives it as a section of Emesis, referring to it the Fabrician species.
1816. Hübn., Verz. 22: restricts it to Thisbe (perdita), and Lysippus (Lysippe).

The latter, therefore, is type, as stated by Crotch (Cist. Ent. 1872). Most authors have considered Melibœus as typical, as would have been the case but for Hübner, Hoffman (Wied. Zöol. Mag. I. ii. 97) specifying only Melibœus of the species given by Fabricius. But the genus is preoccupied in Mollusks (Lam. 1805). See Riodina and Ancyluris.

403. ERYNNIS.

1801. Schrank, Faun. Boica, ii. i. 157: alceæ (malvæ), malvæ (Fritillum), Tages, comma, Thaumas (linea), Morpheus (Speculum).
1820. Oken, Naturg. f. Schulen, 788: alceæ (Malvarum).
1858. Ramb., Cat. Lép. Andal. 83:* Tages (Cervantes), Marloyi.

---

* Rambur also in another work (Faune Ent. Andal.) restricts Erynnis to Tages (Cervantes). The portion of the work containing this (p. 310) may have been printed as early as 1840, but does not appear to have been issued before 1870, judging from the memoranda attached to the copy in the Library of the Entomological Society of France. See also Staudinger's Catalogue, 1871, p. xxx.

1861. Staud., Cat. Lep. Eur. 15: uses it for several species, including Tages.
1871. Kirb., Syn. Cat. 610: employs it for alceæ and others not of Schrank's list. See also Journ. Linn. Soc. Zoöl. x. 498.
1872. Scudd., Syst. Rev. 50: specifies Tages as type.

Neither alceæ nor Tages can, however, be taken as type, for both were previously eliminated (see Urbanus and Thanaos); malvæ was already type of Hesperia in 1798, Thaumas of Adopæa in 1820, and Morpheus of Heteropterus in 1832; there is nothing left but comma, which virtually became the type of the genus in 1832. This necessitates further changes in Pamphila and Cyclopides. See Auglades.

404. ERYPHANIS.

1870. Boisd., Lép. Guat. 57: **Automedon**. Sole species, and therefore type. Also spelled by Boisduval Euryphanis and Eryphane.

Used by Kirby in same sense.

405. ERYTHIA.

1818. Hübn., Verz. 24: **Labdacus** (Labdaca), Gelanor (Gelanoria), Melaphæa, Teleclus (Cataleuce).

Labdacus may be taken as the type.

406. ESOPTRIA.*

1816. Hübn., Verz. 45: *Bolina* (Alcithoe, Bolina). Sole species, and therefore type.

But this name falls before Apatura, through Hübner's own writings. See also Diadema.

407. ESTHEMOPSIS.

1865. Feld., Reise Novara, 306: **Clonia**. Sole species, and therefore type. See Pseudopheles.

408. ETEONA.*

1848. [Boisd. in] Doubl. List Br. Mus. App. 21: *Tisiphone*. Sole species, and therefore type.
1848. [Ib. in] Westw., Gen. Diurn. Lep., pl. 42: the same.
1850. Ib. in Westw., Gen. Diurn. Lep. 254: the same.

The name is preoccupied by Eteone (Sav., Worms, 1817).

409. EUBAGIS.

1832. Boisd., Voy. Astrol. 70: **Athemon** (Arthemon). Sole species, and therefore type.

Used in the same sense by Doubleday and Felder.

410. EUCHARIA.*

1870. Boisd., Lép. Guat. 14: Ganymedes, imperialis, regalis.

The name is preoccupied in Lepidoptera (Hübn., 1816) and in Arachnids (Koch, 1835). See also Evenus, Arcas, and Endymion.

411. EUCHEIRA.*

1834. Westw., Trans. Ent. Soc. Lond. i. 44: *socialis*. Sole species, and therefore type, as indicated by Butler.

The name is preoccupied by Eucheirus (Dej., Col. 1833).

412. EUCHLOE.

1816. Hübn., Verz. 94: Ausonia (Belia, Ausonia), Tagis, Genutia (Midea), **cardamines**, Eupheno, Eucharis (Cœneos).
1841. Westw., Brit. Butt. 30: employs it for cardamines only, and therefore this must be taken as the type. Stephens (1850) and Kirby (1858) make the same use of it.
1872. Scudd., Syst. Rev. 42: specifies Genutia (Midea) as the type, but erroneously. See also Kirby, Zoöl. Rec. 1872, 339. See Anthocharis.

413. EUDÆMON.*

1820. Dalm. in Billb., Enum. Ins. 76: Midamus (Midamus, Claudius, Mulciber), similis, Panope, Clytia (dissimilis), assimilis, Plexippus (Plexippus, Hegesippus), Erippus, Chrysippus (Chrysippus, Alcippus).

The name is preoccupied by Eudæmonia (Hübn., Lep. 1816).

414. EUDAMUS.

1832–33. Swains., Ill. ii. 48: Chalco (Agesilaus), Brachius (Doryssus), **Proteus**. Proteus designated as type.
1833–34. Boisd.-LeC., Lép. Am. Sept., pl. 69: use it for Proteus and many others.
1869. Butl., Cat. Fabr. Lep. 260: uses it, but places Proteus elsewhere.
1870. Ib., Ent. Monthl. Mag. vii. 56: employs it for others than Proteus, and places Proteus elsewhere. See Goniurus.

415. EUEIDES.

1816. Hübn., Verz. 11: **Dianasa**, Halia, Pasinuntia, Eucoma, Mneme, Numata (Pione), Harmonia.
1844. Doubl., List Br. Mus. 57: uses it for Dianasa and an unnamed species.

1848. Ib., Gen. Diurn. Lep. 145: divides the group into two sections, the second containing Dianasa.
1861. Feld., Neues Lep. 6: makes a similar division and reference.
1862. Bates, Linn. Trans. xxiii. 562: uses it for eight species, none of them the originals of Hübner.

Dianasa must be considered as the type.

416. EUERYCINA.

1849. Saund., Trans. Ent. Soc. Lond. [2] v. 97: **Calphurnia.** Sole species, and therefore type.

Thus used by Bates and Kirby. See Rodinia.

417. EUGLYPHUS.*

1820. Dalm. in Billb., Enum. Ins. 80: *Chiron.*

The name is preoccupied by Euglyphis (Hübn., Lep. 1816). See Marius and Megalura.

418. EUGONIA.

1816. Hübn., Verz. 36: c. aureum (Angelica), Polynice, vau. album (v. album), **Polychloros** (Polychloros, Pyrrhomelæna), urticæ, Charonia, Antiopa.
1850. Steph., Cat. Brit. Lep. 12: uses it for Antiopa, Polychloros, and urticæ.
1873. Grote, Can. Ent. v. 144: says that c. aureum (Angelica) is the type; but, owing to the limitation of Stephens, that is impossible.

Polychloros may be taken as the type.

419. EUGRAPHIS.*

1820. Dalm. in Billb., Enum. Ins. 75: *Polyxena* (Hypsipyle). Sole species, and therefore type.

Preoccupied through Eugraphe (Hübn., Lep. 1816). See Thais and Zerynthia.

420. EULACEURA.

1871. Butl., Proc. Zoöl. Soc. Lond. 726: **Osteria.** Sole species and designated type.

421. EULEPIS.

1820. Dalm. in Billb., Enum. Ins. 80: **Athamas.** Sole species, and therefore type.

422. EUMÆUS.

1816. Hübn., Verz. 67: **Minyas.** Sole species, and therefore type.

Subsequently used by various authors in the same sense.

1837–47. Gey. in Hübn., Exot. Schmett. [Eumæa] iii.: uses it for Debora.
    See also Eumenia.

423. EUMENIA.*

1823. God., Encycl. méth. 826: *Minyas* (Toxea). Sole species, and therefore type.
1836. Boisd., Spec. gén. 5 C.: the same species.
    But the name must fall before Eumæus (q. v.). Godart must have borrowed from Hübner in this case, as Eumæus must have been published by 1818, and two such similar names could not have been proposed independently for the same insect.

424. EUMENIS.

1816. Hübn., Verz. 58: Antonoe, Aello, **Semele**, Celimene (Tarpeja).
1850. Steph., Cat. Brit. Lep. 7: uses it for Semele, which therefore becomes the type.
1858. Kirb., List Brit. Rhop.: employs it for Semele and Briseis.

425. EUMESIA.*

1867. Feld., Reise Novara, 504: *semiargentea*. Sole species, and therefore type, as stated by Butler.
    But the name is preoccupied, through Eumesius (Westw., Hym. 1840).

426. EUNICA.

1816. Hübn., Verz. 61: Anna, **Monima**.
1849. Doubl., Gen. Diurn. Lep. 222: employs it for a number of species, including both of Hübner's.
    Used in a similar sense by Felder, Butler, and Kirby. Monima may be taken as the type.

427. EUNOGYRA.

1851. Westw., Gen. Diurn. Lep. 463: **Satyrus**. Sole species, and therefore type.
    Used in same sense by Bates and Kirby.

428. EUOMMA.*

1867. Feld., Reise Novara, 425: angustatum. Sole species mentioned.
    The name is proposed, however, to take the place of Jæra (q. v.) preoccupied, with two species of which, Opis and Crithea, angustatum is congeneric; and therefore one of these two must be taken as the type. But Euomma is itself preoccupied, as Mr. Kirby has pointed out to me, in Coleoptera (Boh. 1858). See Catuna.

429. EUPALAMIDES.*

1816. Hübn., Verz. 101: *Dædalus*. Sole species, and therefore type.
It is not a butterfly.

430. EUPHÆDRA.

1816. Hübn., Verz. 39: Themis, **Cyparissa**, Ceres.
1871. Kirb., Syn. Cat. 247: uses it for the same and many others.
Cyparissa may be taken as the type.

431. EUPHŒADES.

1816. Hübn., Verz. 83: Glaucus, **Troilus**, Palamedes (Chalcas), Polyxenes (Asterius).
1872. Scudd., Syst. Rev. 44: specifies Glaucus as type.
Glaucus, however, cannot be taken as type, being the necessary type of Jasoniades. Troilus may therefore be chosen. See Pterourus.

432. EUPHYDRYAS.

1872. Scudd., Syst. Rev. 27: **Phaeton**. Sole species and designated type.

433. EUPHYES.

1872. Scudd., Syst. Rev. 69: **Metacomet**, vestris, singularis, Osyka, verna. Metacomet specified as type.

434. EUPLŒA.

1807. Fabr., Ill. Mag. vi. 280: Plexippus, similis, Core (Corus).
1816. Ochs., Schmett. Eur. iv. 15: uses it for Chrysippus.
1816. Hübn., Verz. 15: employs it for a number of forms, including only Plexippus of the Fabrician species.
But Plexippus is the type of Danaida, so that this cannot stand. Core would have to be taken as the type (for it is in this sense that it has been used by subsequent authors, such as Boisduval, who invariably spells it Euplæa, Doubleday, Herrich-Schaeffer, and Kirby), were it not that previous to these writers Hübner, in 1816, had taken Core and a close ally to form his Crastia (q. v.); hence similis, which is generically distinct from Plexippus, must be taken as the type.
1872. Crotch, Cist. Ent. i. 66: says that Leucostictos (Eunice) is the type, through Boisduval in 1832; but it is not one of the species mentioned by Fabricius.

435. EUPTOIETA.

1848. Doubl., Gen. Diurn. Lep. 168: Hegesia, **Claudia**.
Used in same sense by Felder and Kirby.
1872. Scudd., Syst. Rev. 22: specifies Claudia as the type.

436. EUPTYCHIA.

1816. Hübn., Verz. 54 : **Herse**, Penelope (Clarissa), Hesione (Ocyrrhoe), Ocypete, Cephus (Lisidice), Hermes (Hermessa), Mollina (Molina), Lydia, Junia, Libye, Tolumnia, Chloris (Chlorimene), Arnæa (Ebusa), Myncea, and a MS. species.
1844. Doubl., List Br. Mus. 122 : uses it for four species, of which three are Hübner's; viz., Hesione (Cissia), Arnæa (Ebusa), and Tolumnia.
1851. Westw., Gen. Diurn. Lep. 372 : employs it for the, same and others, including Herse.
1868. Butl., Ent. Monthl. Mag. iv. 194 : specifies Herse as type.
1871. Kirb., Syn. Cat. 47 : uses it for all of Hübner's species and others.

As Herse appears to be strictly congeneric with Tolumnia, it can be accepted as the type.

437. EURALIA.

1850. Westw., Gen. Diurn. Lep. 281 : **dubius** (dubia), Anthedon.
1861. Feld., Neues Lep. 25 [Eucalia] : the same.

Dubius may be considered the type.

438. EUREMA.

1816. Hübn., Verz. 96 : **Delia** (Demoditas), Elathea, Sylvia (Eudoxia), Agave (Jodutta, Phiale), albula (Nise), Hecabe.
1844. [Boisd. in] Doubl., List Br. Mus. 83 : uses it for Lethe and Zabulina, which have nothing to do with Hübner's genus. See Hypanartia.
1848. Boisd. in Doubl., Gen. Diurn. Lep. 176 : makes the same general use of it.
1850. Steph., Cat. Brit. Lep. 252 : employs it for Philodice and Palæno (Palæno, Europome), which is nearer the Hübnerian mark, but still erroneous.
1861. Feld., Neues Lep. 12 : uses it in Doubleday's sense.
1870. Boisd., Lép. Guat. 39 : does the same.
1870. Butl., Cist. Ent. i. 35 : designates the type as Delia.
1872. Scudd., Syst. Rev. 39 : the same.

Should it be written Heurema?

439. EURHINIA.

1867. Feld., Reise Novara, 405 : Polynice (Elpinice, Megalonice, Stratonice). Sole species; the name Eurhinia, however, is evidently given to supplant Rhinopalpa (q. v.) preoccupied, of which **fulva** was the type.

Fulva, therefore, and not Polynice, must be taken as the type of Eurhinia. This name, however, is certainly very close to Eurhina (Fitz., Rep. 1843) and Eurhinus (Kirb., Col. 1817).

440. EURIPHENE.

1847. Boisd., Voy. Deleg. ii. 592: **cærulea.** Sole species, and therefore type.

Used in same sense (but spelled Euryphene) by Westwood, Felder, and Kirby.

441. EURIPUS.

1848. Doubl., Gen. Diurn. Lep., pl. 41: **Halitherses.** Sole species, and therefore type.
1850. Westw., Gen. Diurn. Lep. 293: Halitherses, consimilis (Hallirothrius).
1861. Feld., Neues Lep. 26: uses it in the same sense.

The name is rather near to Eurrhypis (Hübn., Lep. 1816).

442. EURYADES.

1864. Feld., Spec. Ins. 39: Duponchelii, **Corethrus.**

Used with exactly the same limitation by Herrich-Schaeffer and Kirby. Inasmuch as the Felders remark that they had not seen the first species, Corethrus may be taken as the type.

443. EURYBIA.

{ 1816. Hübn., Verz. 17: **Nicæus** (Nicæa), Halimede, Dardus (Upis).
{ 1818. Ill., Wied. Zoöl. Mag. i. ii. 100: Nicæus, Halimede, Lamia.

With the exception of the last species in each case, the usage of Hübner and Illiger is identical. Recalling the statements made in the introduction to this essay, there can be no question that this name should be credited to Illiger.

1819. God., Encycl. méth. 459: uses it for Carolina, Nicæus, and Dardus.
1832. Guér., Iconogr., pl. 80, fig. 4 [Erybia]: carolina.

Whether Illiger or Hübner have priority, Nicæus, by Godart's usage, must become the type.

444. EURYCUS.

1836. Boisd., Spec. gén. 391: **Cressida** (Cressida, Harmonia). Sole species, and therefore type.

In this sense it has been used by all subsequent authors. See Cressida.

445. Eurygona.

1836. Boisd., Spec. gén., pl. 3, 5 C.: Crotopus (Midas) [larva and pupa only], **Phædica.**
1847. Doubl., List Br. Mus. 5: employs it for several species, but for neither of the above.
1851. Westw., Gen. Diurn. Lep. 437: employs it for many species, including both of Boisduval's and some of Doubleday's.
1867. Bates, Journ. Linn. Soc. Lond. ix. 420: employs it for more than sixty species, including both of Boisduval's.
Phædica may be taken as the type.

446. Eurymus.

1829. Swains. in Horsf., Descr. Cat. Lep. E. Ind. Mus. 129, 134: **Hyale.** Sole species given, and designated type. Said by Horsfield to have been so given him by Swainson about eight years previously.
1832–33. Swains., Zoöl. Ill. ii. 60, 70: Philodice (Philodice, Europome). See Colias.

447. Eurytela.

1833. Boisd., Ann. Mus. Hist. Nat. 202: Horsfieldii (Horsfieldii, Stephensii), **Dryope.**
1844. Doubl., List Br. Mus. 145: uses it for Dryope and two others.
Dryope is therefore the type.

448. Eurytides.

1822–26. Hübn., Exot. Schmett. ii.: **Dolicaon,** Iphitas.
Dolicaon may be taken as the type.

449. Euschemon.[*]

1846. Doubl. in Stoke's Austr. i. App. 513: *Rafflesia*. Sole species, and therefore type.
Preoccupied, through Euschema (Hübn., Lep. 1816).

450. Euselasia.

1816. Hübn., Verz. 24: Crotopus (Crotope), Hygenius (Hygenia), Orfita (Orsita), Arbas (Arbassa), Sabinus (Tenage), Euriteus (Cynira[*]), **Gelon** (Gelæna), Teleclus (Telecta).
1871. Kirb., Syn. Cat. 294: uses it for all the above, and many more.
Gelon may be selected as type.

---

[*] This name is an accidental error of Hübner's in copying from Cramer.

451. EUTERPE.*

1832–33. Swains., Ill. ii. 74: **Tereas** (Terea). Sole species, and therefore type, as stated by Butler.

> Used in same sense by Boisduval, Doubleday, and Herrich-Schaeffer. According to Kirby (Syn. Cat.), this name is preoccupied (but not in zoölogy): it falls, however, before Archonias (q. v.).

452. EUTHALIA.

1816. Hübn., Verz. 41: **Lubentina**, Adonia
1871. Kirb., Syn. Cat. 252: uses it for the above and others.

> Lubentina may be taken as the type.

453. EUTHYMUS.*

1872. Scudd., Syst. Rev. 56: *Phylæus*. Sole species and designated type.

> The name falls before Hylephila.

454. EUTRESIS.

1847. Doubl., Gen. Diurn. Lep. 111: **Hypereia**. Sole species, and therefore type.
1871. Kirb., Syn. Cat. 19: the same.

455. EUXANTHE.

1816. Hübn., Verz. 39: **Eurinome**. Sole species, and therefore type.
1871. Kirb., Syn. Cat. 228: the same and another.

> See Anthora and Godartia.

456. EVENUS.

1816. Hübn., Verz. 78: **regalis** (Endymion), Ganymedes. Regalis may be taken as the type.

> See Eucharia, Endymion, and Arcas.

457. EVERES.

1816. Hübn., Verz. 69: **Argiades** (Amyntas, Polysperchon). Sole species, and therefore type, as indicated by Scudder (Syst. Rev.).

458. EVONYME.

1816. Hübn., Verz. 61: **Amelia**, Sophonisba.

> This generic name has never since been employed. Amelia may be taken as the type.

459. Fabius.*

1837. Dunc., For. Butt. 167: *Hippona*. Sole species, and therefore type.

But as Fabius is one of the synonymes of Hippona, the name falls. See Consul, Helicodes, and Protogonius.

460. Faunia.*

1847. Poey, Mem. Soc. Econ. Habana, [2] iii. 178: *Orphise* (Orphisa). Sole species, and therefore type.

The details of Herrich-Schaeffer's reference (Schmett. Cuba, 5) are erroneous.

1867. Feld., Reise Novara, 406: Olympias, Persephone, Tithonia, Vemesia, Pomona, Araucana.

The name is preoccupied in Diptera (Rob.-Desv., 1830), and very near to Faunis (Hübn., Lep. 1816) and Faunus (Montf., Moll. 1810).

461. Faunis.*

1816. Hübn., Verz. 55: Eumeus (Eumea), Echo.

Preoccupied through Faunus (Montf., Moll. 1810). See Clerome.

462. Faunula.

1867. Feld., Reise Novara, 488: **Leucoglene**. Sole species, and therefore type, as stated by Butler.

463. Feniseca.

1869. Grote, Trans. Amer. Ent. Soc. ii. 308: **Tarquinius**, Porsenna. Tarquinius specified as type, as stated by Scudder.

464. Festivus.*

1872. Crotch, Cist. Ent. i. 62: refers this name to Fabricius, and says that Latreille (1805) fixed the type as Plexippus; but see our introductory remarks.

465. Ganoris.*

1816. Dalm., Vetensk. Acad. Handl. xxxvii. 61, 86: I. cratægi, *brassicæ*, rapæ, napi, Daplidice, cardamines, sinapis; II. Hyale, Palæno, rhamni. Brassicæ is specified as the type.

1872. Scudd., Syst. Rev. 41: designates rapæ as type, but erroneously. See Pieris.

Brassicæ having previously been made the type of Mancipium, this name falls, and cannot be employed again. See also Pontia.

466. GANYRA.

1820. Dalm. in Billb., Enum. Ins. 76: Leucippe, Croceus (Edusa), Hyale, Palæno (Paleno), Hecabe, Nise, Proterpia, Elathea, albula, Monuste, Pyranthe (Gnoma, Minna), **Amaryllis**, Crocale (Alcmeone), Scylla, Argante (Hersilia), Eubule, Trite, and a number of MS. species.

Amaryllis may be taken as the type.

467. GEGENES.

1816. Hübn., Verz. 107: **Pygmæus** and two MS. species. Pygmæus must therefore be considered the type.

1870. Butl., Ent. Monthl. Mag. vii. 93: specifies Pygmœus (Pygmæa) as type.

468. GEITONEURA.*

1867. Butl., Ann. Mag. Nat. Hist. [3] xix. 164: *Klugii*, Achanta.
1868. Ib., Ent. Monthl. Mag. iv. 196; and Cat. Sat. 166: specifies Klugii as type.

The name falls before Xenica (q. v.).

469. GERYDUS.

1836. Boisd., Spec. gén., pl. 7 C.: **Symethus**. Sole species, and therefore type.

Used for the same species by Doubleday (List). See Symetha and Miletus.

470. GLAUCOPSYCHE.

1872. Scudd., Syst. Rev. 33: **Lygdamus**, Pembina. Lygdamus specified as type.

471. GLOBICEPS.*

1869. Feld., Pet. Nouv. Ent. i. viii.: *paradoxa*. Sole species, and therefore type.

The generic name is preoccupied in Hemiptera (Lep.-Serv. 1825). See Pseudopontia and Gonophlebia.

472. GLYCESTHA.

1820. - Dalm. in Billb., Enum. Ins. 76: cratœgi, Hyparete (Hyparite), Pasithoe, **Java** (Coronea).

Java may be taken as the type.

473. GNATHOTRICHE.

1862. - Feld., Wien. Ent. Monatschr. vi. 420, note: **exclamationis**. Sole species, and therefore type.

474. GNESIA.

1848. Doubl., Gen. Diurn. Lep. 141: Medea, Zetes (Menippe, Zetes), Persephone, Egina, Perenna, **Circeis**.
   Circeis may be taken as the type.

475. GNOPHODES.

1849. Doubl., Gen. Diurn. Lep., pl. 61: **Parmeno.** Sole species, and therefore type.
1851. Westw., Gen. Diurn. Lep., 363: Parmeno, Chelys (Morpena).
1868. Butl., Ent. Monthl. Mag. iv. 194: designates Parmeno as type.

476. GODARTIA.*

1842. Luc., Ann. Soc. Ent. Fr. xi. 297: *madagascariensis*. Sole species, and therefore type.
1850. Westw., Gen. Diurn. Lep. 282; Eurinome, madagascariensis.
   The name is very close to Gœdartia (Boie., Hym. 1841), though named after another person. It is, however, synonymous with Euxanthe, and must fall before it. See also Anthora.

477. GODYRIS.*

1870. Boisd., Lép. Guat. 33: *Duillia*. Sole species, and therefore type.
   It falls before Hymenitis.

478. GONEPTERYX.*

1815. Leach, Edinb. Encycl. 716: *rhamni*. Sole species, and therefore type.
1827. Curtis, Brit. Ent. pl. 173: designates rhamni as the type.
1827. Steph., Ill. Brit. Ent. Haust. 8: uses it for rhamni only.
1840. Westw., Gen. Syn. 87 [Goniapteryx]: rhamni given as type.
1847. Doubl., Gen. Diurn. Lep. 69: uses it for many species with rhamni.
1853. Wallengr., Rhop. Scand. 145 [Goniopteryx]: rhamni.
1870. Butl., Cist. Ent. i. 35, 45: specifies rhamni as type.
   The generic name falls, however, before Colias (q. v.). See also Gonoptera and Earina.

479. GONILOBA.

1852. Westw., Gen. Diurn. Lep. 512: Creteus, Celænus, Vespasius (Cassander), Parmenides, Bixæ, Apastus, Aulestes, Hylaspes, Pervivax, Scipio, Mercatus (fulgerator), Talus,

Corytas (Pyramus), Amyntas (Savignyi), Phidon*
(Phedon), Cometes, Schonherri, Idas (Mercurius), Tityrus, Yuccae,* Olynthus,* Exadeus, Epitus,* Evadnes*
(Evadne), Pomus (Comus), Brino,* dubius, Anaphus,
Orchamus,* Pompeius (Archalaus), Ericus, Chromus,
Alexis, Euribates, Salatis, Muretus, Ramusis, Midas
(Rhetus), Ethlius* (Chemnis, Ethlius), Hesus, Corydon
(Coridon), Lucasii (Lucas), Antoninus,* Salius,* Nyctelius, Dalmanni,* Basochesii, Fischeri,* Lesueuri, Bonfilius, Dan, Sergestus, Feisthamelii, Sabadius, Japetus
(Nepos), Phineus, Lucretius, Minos, **Xanthaphes** * (Xanthoptes), Aristoteles, Justinianus, Lafrenayii, Fantasos,
Helops, Phocus, Avitus, Crinisus, Ebusus, Psecas, Alcmon, Artemides, Zestos, Bathyllus (Bethyllus), Astylos,*
Broteas,* Corytas, Vulpinus, Olenus,* Nicias, Godartii.

1869. Herr.-Schaeff., Prodr. iii. 69 : gives a large number of species, including those of the above list which are followed by an asterisk.

1870. Butl., Ent. Monthl. Mag. vii. 56 : uses it for Tityrus, Exadeus, and others not mentioned by Westwood.

None of Butler's species being congeneric with those employed in this group by Herrich-Schaeffer, Butler's action has no effect whatever upon the determination of a type. Of Westwood's species mentioned by Herrich-Schaeffer, Phidon, Ethlius, and Olenus are types of other genera. This group may be confined to Xanthaphes and allies. See Niconiades.

480. GONIURUS.

1816. Hübn., Verz. 104: Simplicius, Dorantes, Brachius (Brachyus), Cœlus, Catillus, Proteus, Tarchon, Eudoxus, Orion.

1852. Westw., Gen. Diurn. Lep. 510 [Goniuris]: employs' it for a dozen species, including all of the above.

1869. Butl., Cat. Fabr. Lep. 259 [Goniuris]: employs it for Proteus only.

1870. Ib., Ent. Monthl. Mag. vii. 56: specifies Simplicius as type.

But neither Proteus nor Simplicius can be taken as the type, since they are congeneric, and Proteus has been taken as the type of Eudamus, carrying with it most of Hübner's Goniuri. Cœlus may be taken as the type.

---

* See the succeeding entry.

481. GONOPHLEBIA.

1870 (Aug.). Feld., Pet. Nouv. Ent. 95: **paradoxa**. Sole species, and therefore type.

 Proposed to supplant Globiceps, preoccupied. Is it a butterfly? See also Pseudopontia.

482. GONOPTERA.*

1820. Dalm. in Billb., Enum. Ins. 76 [Gonrptera]: rhamni (rhemni), Cleopatra.

 Falls before Colias (q. v.). See also Gonepteryx and Earina.

483. GONOPTERIS.*

1832. Gey. in Hübn., Zutr. iv. 34: *Pergæa*. Sole species, and therefore type.

 The name is preoccupied through Gonoptera (Dalm., Lep. 1820, and Latr., Lep. 1825).

484. GORGO.*

1816. Hübn., Verz. 64: Ceto, Medusa, Œme (Psodea, Œme).

 The name falls before Erebia. See also Marica, Syngea, Phorcis, Epigea, and Oreina.

485. GRAPHIUM.*

1777. Scop., Introd. 433: Medon and an immense number of wholly disconnected species, arranged in eight divisions.

 None of these divisions (when they contain more than a single species) are homogeneous. Take, for example, the second, which among others contains Sarpedon [Papilionides], Mneme [Tribuni], populi [Archontes], and Clio [Hamadryades]; or the fourth with these: Euterpe [Stalachtis], Charithonia [a Heliconian], and Venilia [Athyma]. Every one of the families are represented. The genus must therefore be dropped as thoroughly discreditable to the author, even at the early time it was established.

486. GRAPTA.*

1837. Kirb., Faun. Bor. Amer. 292: *Progne* (c. argenteum). Sole species, and therefore type.

1848. Doubl., Gen. Diurn. Lep. 195: employs it for a number of species, including the above.

 He gives Polygonia as a synonyme, but evidently at one time intended to use it in preference to Grapta, since he elsewhere in the text (p. 199) refers to this genus as Polygonia.

1861. Feld., Neues Lep. 12: uses it in the same sense.

 It has been elsewhere very generally adopted, but is synonymous with Polygonia, and must fall before it. See also Comma.

487. GYNŒCIA.*

1844. Doubl., List Br. Mus. 88: *Dirce*. Sole species, and therefore type.

It has been used in the same sense by Westwood, Kirby, and Felder, the last of whom spells it Gynæcia; but it falls before Colobura (q.v.)

488. GYROCHEILUS.

1867. Butl., Ann. Mag. Nat. Hist. [3] xx. 267: **Patrobas**. Sole species and designated type.

489. HADES.

1851. Westw., Gen. Diurn. Lep. 435: **Noctula**. Sole species, and therefore type.

Used for the same species by Bates and Kirby. See Moritzia.

490. HADOTHERA.*

1820. Billb., Enum. Ins. 80: proposed, without reason, to supplant Danis. No species are referred to it.

491. HÆMATERA.

1848. Doubl., Gen. Diurn. Lep., pl. 30: **Thysbe**. Sole species, and therefore type.
1849. Ib., ib. 231: Pyramus, Thysbe.

Subsequently used in the same sense by Felder and Kirby.

492. HÆMONIDES.

1816. Hübn., Verz. 101: **Cronis**.* Sole species, and therefore type.

493. HÆTERA.

1807. Fabr., Ill. Mag. vi. 284: **Piera**, diaphanus.
1820. Billb., Enum. Ins. 77: without apparent reason, but much according to his wont, changes the name to Pselna.
1836. Boisd., Spec. gén., pl. 9 B.: Piera is figured, and therefore this must be taken as type. It has been used by subsequent authors in the same sense.
1868. Butl., Ent. Monthl. Mag. iv. 195: designates Piera as type.
1872. Crotch, Cist. Ent. i. 66: says that Piera is type through Doubleday in 1846, overlooking Boisduval's action.

494. HAMADRYAS.

1806. Hübn., Tent. 1: **Io**. Sole species, and therefore type.
1832. Boisd., Astrol. 91: employs it for Zoilus and Assarica (Assaricus), which have no connection with Hübner's group.

Since used by many authors in the later sense. See also Inachis.

---

* See note, p. 293.

495. HAMANUMIDA.

1816. Hübn., Verz. 18: Veronica, **Dædalus** (Meleagris), Flegyas (Allica), Actoris (Actoria), Thasus (Thase), Ceneus (Lusia).
1871. Kirb., Syn. Cat. 249: employs it for Dædalus only, which therefore becomes type.

See also Canopus.

496. HAMEARIS.

1816. Hübn., Verz. 19: Abaris (Abarissa), **Epulus** (Epule), Lucina.
1830. Curtis, Brit. Ent., pl. 316: designates Lucina as the type; but Stephens's action in the previous year, in founding the genus Nemeobius, renders this nugatory.
1840. Westw., Gen. Syn. 88: specifies Lucina as type.
1867. Bates, Journ. Linn. Soc. Lond. ix. 447: employs it for several species, including Epulus only of Hübner's species, and this therefore becomes the type.

497. HAMES.*

1851. Boisd. MS. in Westw., Gen. Diurn. Lep. 366: mentioned by Westwood as synonymous with Cærois, but not otherwise referred to by any writer. Boisduval himself has never mentioned it, and no species have been referred to it.

498. HARMA.*

1848. Boisd. in Doubl., Gen. Diurn. Lep., pl. 40: *Theobene*. Sole species, and therefore type.
1850. "Doubl." [but erroneously] in Westw., Gen. Diurn. Lep. 287: Theobene and others.
1861. Feld., Neues Lep. 33: divides the group into three sections, the first comprising Theobene.

The name is preoccupied by Arma (Hahn, Hemipt. 1833).

499. HEBOMOIA.

1816. Hübn., Verz. 96: **Glaucippe**, Leucippe.
1847. Doubl., Gen. Diurn. Lep. 62: the same.
1870. Butl., Cist. Ent. i. 37, 48: specifies Glaucippe as type.

See Iphias.

500. HECAERGE.*

1816. Ochs., Schmett. Eur. iv. 32: *celtis*. Sole species, and therefore type.

1816. Hübn., Verz. 100: Carinenta, celtis.

Besides the reasons given in the introduction for believing that Hübner's Verzeichniss did not appear until after 1816, which alone would be enough to give Ochsenheimer the preference in this case, Ochsenheimer's preface is dated in March and Hübner's in September. One must have borrowed from the other. It is beyond credence that both should have coined the same generic word for the same insects, unless there were some special significance in the name, as there is not. Hübner's genus was defined (briefly), while Ochsenheimer's was not; but the latter author gives a reason (an insufficient one) for changing the name of the earlier Libythea, just as he does in the case of Charaxes; and there can therefore be little doubt that the genus is to be credited to Ochsenheimer. *In that case*, the genus cannot stand, for celtis (which is generically distinct from Carinenta) had already been taken as the type of Libythea. See also Hypatus.

501. HECALENE.*

1844. [Boisd. in] Doubl., List Br. Mus. 112: *Clytemnestra*. Sole species, and therefore type.

But this name must fall before Hypna (q. v.), as pointed out by Westwood, in the Genera of Diurnal Lepidoptera, where Hecalene is credited to Boisduval.

502. HECTORIDES.

1822. Hübn., Index: **Agavus**, Brunichus.
1822–26. Ib., Exot. Schmett. ii.: Lysithous, Ascanius.
1825. Ib., Zutr. iii. 25: Proneus.

The choice of type must, of course, lie between Agavus and Brunichus, and Agavus may be selected.

503. HEDONE.*

1872. Scudd., Syst. Rev. 58: *Brettus*, Præceps, Coscinia, Orono, Ætna. Brettus specified as type.

It falls before Thymelicus (q. v.).

504. HELCYRA.

1860. Feld., Sitzungsb. Acad. Wien, xl. 450: **Chionippe**. Sole species, and therefore type.
1861. Ib., Neues Lep. 37, 44: the same.

505. HELIAS.*

1807. Fabr., Ill. Mag. vi. 287: no insects cited, excepting an unnamed MS. species.

The description is also entirely insufficient to give any clew to what Fabricius may have had in mind, and hence the name must be dropped.

1820. Billb., Enum. Ins. 80: proposes, for no reason, to change the name to Achna; he also mentions no species.
1867. Feld., Reise Novara, 531: uses it for seven new species, allied to Busiris and others.
1870. Butl., Ent. Monthl. Mag. vii. 98: specifies phalænoides as type.
1871. Kirb., Syn. Cat. 634: follows Butler, but questions whether it is used in the Fabrician sense.

See Achlyodes.

506. HELICODES.*

1844. [Boisd. in] Doubl., List Br. Mus. 112: *Hippona*. Sole species, and therefore type.
1850. Boisd. MS. in Westw., Gen. Diurn. Lep. 318: Westwood gives it as a synonyme of Protogonius.
1870. Boisd., Lép. Guat. 49: claims it as his own, placing the same species in it.

It falls, however, through Consul. See also Fabius and Protogonius.

507. HELICONIUS.

1805. Latr., Sonn. Buff. xiv. 108: **Antiochus** (Anthioca). Sole species, and therefore type.
1809. Ib., Gen. Crust. et Ins. iv. 200: divides the group in two sections, but does not specify Antiochus in either.
1810. Ib., Consid. 440: specifies Polymnia and Horta as types (!), these being the first species of each section in his previous work.
1815. Oken, Lehrb. i. 725: treats it as Latreille in his later works.
1817. Latr., Cuv. Règne Anim. iii. 549: employs it for a number of species, but Antiochus is not mentioned.
1823. Hübn., Zutr. ii. 31 [Heliconia]: employs it for Lansdorfii (Langsdorfii), which has nothing to do with the Fabrician genus.
1836. Boisd., Spec. gén., pl. 7 B. [Heliconia]: figures Dœta.

It is subsequently used for species allied to Anthioca by Doubleday, Bates, and others.

1872. Crotch, Cist. Ent. i. 60: refers the genus back to Linné [Heliconii], but erroneously, and says the type was fixed by Lamarck in 1801 as Psidii.

But Lamarck at this time only divided the genus Papilio into sections, giving them the Linnean names in the plural form, and specified Psidii as an example of Heliconii.

508. HELICOPIS.

1807. Fabr., Ill. Mag. vi. 285: **Cupido**, Acis (Gnidus).
1815. Oken, Lehrb. i. 722: uses it as a section of Emesis, referring to it the same species, together with Endymion.
1816. Hübn., Verz. 22: employs it for Cecilia (Cicilia), which is only distantly related to the Fabrician types.
1818. Hoffm. in Wied., Zoöl. Mag. i. ii. 98: refers the Fabrician species to it.
1836. Boisd., Spec. gén., pl. 3 A.: gives an illustration of Cupido, which therefore becomes type.

It has been used in same sense by later authors. See also Erotion and Hexuopteris.

509. HELIOCHLÆNA.*

1822. Hübn., Index: *Leucosia*. Sole species, and therefore type.

The name falls before Peplia. See Desmozona and Nymphidium.

510. HELIOCHROMA.

1869. Butl., Cist. Ent. i. 15: **idiotica**. Sole species, and therefore type, as subsequently indicated by Butler. See p. 293.

511. HELIOPETES.

1820. Billb., Enum. Ins. 81: **Arsalte** (niveus) and a MS. species. Arsalte therefore is the type.

See also Leucoscirtes.

512. HELIOPHORUS.

1832. Gey. in Hübn., Zutr. iv. 40: **Epicles** (Belenus). Sole species, and therefore type.

See also Ilerda.

513. HELIORNIS.*

1820. Dalm. in Billb., Enum. Ins. 79: Laertes (Epistrophus), Menelaus (Menelaus, Nestor), Achilles (Helenor, Achilles).

This name is preoccupied in Birds (Bonn. 1790).

514. HEMEROCHARIS.*

1836. Boisd., Spec. gén. 412: given only as a MS. synonyme of Leptalis by the author himself. It therefore cannot be used in this (or any other) sense.

515. HEMIARGUS.

1816. Hübn., Verz. 69: Bubastus, Parsimon (Celæus), Lysimon (Ubaldus), **Hanno**, Isis (Isarchus), Larydas, and a MS. species.

Hanno may be selected as the type.

516. HEODES.

1816. Dalm., Vetensk. Acad. Handl. xxxvii. 63, 91: Hippothoe (Hippothoe, Chryseis), Virgaureæ, **Phlæas**, Helle, Dorilas (Garbas), rubi.
1820. Billb., Enum. Ins. 80: the same, excepting rubi, and others.
1835. Vill.-Guén., Lép. Eur. 32: Helle, Phlæas, and other coppers.
   Phlæas may be taken as the type. See Lycæna.

517. HERACLIDES.

1816. Hübn., Verz. 83: **Thoas** (Oxilus, Thoas), Menestheus, Pelaus, Demolion (Cresphontes), Phorcas.
   Thoas may be taken as the type. See also Thoas.

518. HERONA.

1848. Doubl., Gen. Diurn. Lep. pl. 41: **Marathus**. Sole species, and therefore type.
1850. Westw., Gen. Diurn. Lep. 293: the same.

519. HERPÆNIA.*

1870. Butl., Cist. Ent. i. 38, 52: *Eriphia* (Tritogenia). Sole species and designated type.
   The name must fall before Picanopteryx.

520. HESPERIA.

1793. Fabr., Ent. Syst. iii. i. 258: established upon all the Rurales and Urbicolæ known to him, three hundred and forty-nine names (231 Rurales, 118 Urbicolæ), the two groups commencing respectively with Cupido and exclamationis among the latter **malvæ**.
1798. Cuv., Tabl. Élém. 592: cites malvæ as an example and the only one. This, therefore, becomes the type, being one of those used by Fabricius.
1807. Fabr., Ill. Mag. vi. 285: employs it for Amor, Helius, Faunus, Vulcanus, Marsyas, Bœtica, Acmon (Æmon), Thysbe, Thetys (Æsopus), and Pretus, all Rurales, to which group, but for Cuvier's action, Hesperia would have to be restricted; as it is, Fabricius's action has no effect.
1810. Latr., Consid. 440: specifies Proteus, malvæ, and Morpheus (Steropes) as types.
1815. Oken., Lehrb. i. 720: employs it for Helle and allies!
1816. Dalm., Vetensk. Acad. Handl. xxxvii. 200: specifies comma as type, but erroneously.

1816. Lam., Hist. Nat. An. sans Vert. iv. 20 : employs it for malvæ and others.
1816. Hübn., Verz. 25 : uses it for various Vestales, following Fabricius' own tardy limitation, although not in precisely the same sense.
1820. Billb., Enum. Ins. 81 : some Urbicolæ, among them malvæ.
1820. Oken, Naturg. f. Schulen, 788 : employs it for some Ephori.
1820-21. Swains., Zoöl. Ill. i. i. 28 : specifies comma as the type, but erroneously.
1833. Curtis, Brit. Ent., pl. 442 : also designates comma as the type.
1837. Sodoffsk., Bull. Mosc. x. 82 : proposes to supplant this name by Symmachia (q. v.).
1840. Ramb., Faun. Ent. Andal. 312 [probably unpublished] : uses it for a number of species, including malvæ (Alveolus).
1852. Westw., Gen. Diurn. Lep. 525 : employs it for a heterogeneous group of Urbicolæ, not including malvæ.
1858. Ramb., Cat. Lép. Andal. 88 : limits it wrongly to Nostrodamus (Nostradamus).
1858. Kirb., Cat. Brit. Rhop. : limits it to comma.
1869. Butl., Cat. Fabr. Lep. 269 : employs it for exclamationis and others, but not for malvæ.
1870. Ib., Ent. Monthl. Mag. vii. 58 : specifies exclamationis as the type, erroneously.*
1870. Kirb., Journ. Linn. Soc. Lond. x. 500 : says that Proteus seems to be Latreille's type, and Alcides that of Fabricius.

---

* Butler (Lep. Exot. 166, note) says of Hesperia : "Fabricius described the genus in his Entomologia Systematica, vol. iii., Gloss. 1, p. 325 (1793), and gave *no type*, but used the following words in his description — ' Antennæ clava elongata, sæpius uncinata.' These words at once fix the type as somewhere amongst the *Hesperiæ urbicolæ* (notwithstanding the fact that, in his Systema Glossatorum, Fabricius refers it to the *rurales*). The *Hesperia* of Cuvier has for its type *H. Malvæ* (as Mr. Crotch has pointed out, Cist. Ent. p. 62) ; but *Pyrgus Malvæ* (of all the *Hesperiæ urbicolæ*) is about the worst to have chosen as the type, for it *does not fit* the Fabrician description. Therefore it is clear that *P. Malvæ* could not have crossed the mind of Fabricius when he penned his description, and *cannot be* his type : later authors have referred the dark-coloured species of *Pamphila* and *Carystus* to *Hesperia*, evidently taking *H. Exclamationis* as the type, it being the first species which he describes under his *urbicolæ* ; but as *H. Exclamationis* turns out to be an *Ismene*, and not, as formerly supposed, a *Pamphila*, I have taken *I. Exclamationis* as the type. The first of the *Hesperiæ Rurales* is a species of the family Erycinidæ."

1871. Ib., Syn. Cat. 611: places a large number of species in the group, including malvæ, but excluding comma, exclamationis, and Proteus, showing that he doubtless considers malvæ as the type.
1872. Crotch, Cist. Ent. i. 62: says that malvæ is the type, through Cuvier, 1799.
1872. Scudd., Syst. Rev. 52: specifies malvæ as the type.

All of the species indicated above under this heading, excepting some of those not specified by name, were placed by Fabricius under Hesperia at its establishment. See Pyrgus, Scelothrix, and Syrichtus.

521. HESPERILLA.

1868. Hewits., Hundr. Hesp. 37: **ornata**, Halyzia, Doubledayi (Dirphia), Donnysa, Peronii (Doclea). Ornata specified as type.
1871. Kirb., Syn. Cat. 622: uses it in the same sense.

See Telesto.

522. HESPEROCHARIS.

1862. Feld., Verh. zoöl.-bot. Gesellsch. Wien, xii. 493: I. **Erota**, Marchalii, Helvia, Nera, Anguitia; II. Gayi. See p. 293.
1867. Herr.-Schaeff., Prodr. ii. 17: uses it in the same sense.
1870. Butl., Cist. Ent. i. 34, 42: designates Erota as type.

523. HESTIA.

1816. Hübn., Verz. 15: similis, assimilis, Idea, **Lynceus** (Lyncea), Ismare, Menelaus (Ephyre), Juventa, Plexippus (Thoe).
1844. Doubl., List Br. Mus. 52: uses it for Idea, Lynceus, and two others not of Hübner's list.
1847. Doubl., Gen. Diurn. Lep. 94: uses it in the same sense.
1871. Kirby, Syn. Cat. 1: follows Doubleday.

Since Lynceus is generically distinct from Idea, it may be taken as the type. See Idea and Nectaria.

524. HESTINA.

1850. Westw., Gen. Diurn. Lep. 281: I. **assimilis**, persimilis, consimilis, Nama; II. Nyctelius, Pimplea?
1861. Feld., Neues Lep. 25: limits it to the first section, which he again divides into two, using assimilis and Nama as the types of the two divisions.
1871. Kirb., Syn. Cat. 227: uses it in the Felderian sense.

Assimilis may be considered as the type.

525. HETEROCHROA.

1836. Boisd., Spec. gén., pl. 4 B.: **Serpa.** Sole species, and therefore type.
1844. Doubl., List Br. Mus. 106: employs it for a great number of species, including Serpa.
1850. Westw., Gen. Diurn. Lep. 276: uses it in the same sense.
1861. Feld., Neues Lep. 28: divides it into two sections.

    According to Kirby (Syn. Cat.), the name is preoccupied; but only in botany, as he informs me by letter.

526. HETERONYMPHA.

1858. Wallengr., K. Vet. Akad. Förhandl. xv. 78: **Merope,** Abeona.
1868. Butl., Ent. Monthl. Mag. iv. 195; and Cat. Sat. 99: specifies Merope as type.

    See also Tisiphone, Hipparchioides, and Xenica.

527. HETEROPSIS.

1850. Boisd. in Westw., Gen. Diurn. Lep. 323: **Drepana.** Sole species, and therefore type.
1871. Kirb., Syn. Cat. 96 (referred to Westwood, not Boisduval): the same.

528. HETEROPTERUS.

1806. Dum., Zoöl. Anal. 271: no species mentioned; he refers to it all Urbicolæ with wings *croisées,* the rest being grouped under Hesperia.
1823. Ib., Consid. 222, pl. 41: **Morpheus** given as an example. It is therefore the type.
1832. Dup., Pap. France, Diurn. Suppl. 413: employs it for Morpheus (Aracinthus), Palæmon (Paniscus), and sylvius.
1840. Ramb., Faun. Ent. Andal. 305 [unpublished?]: refers to it lineola and four other species no more nearly allied to Morpheus than it is. So also in his Faun. Andal.
1853. Wallengr., Scand. Rhop. 250: limits it to sylvius.
1858. Ib., Rhop. Caffr. 46: uses it, more correctly, for Metis and Willenii.
1870. Kirb., Journ. Linn. Soc. Lond. x. 500: says that Morpheus (Speculum) is the type. See also Cyclopides.

529. HEUREMA.*

1867. Herr.-Schaeff., Prodr. ii. 8: *impura.* Sole species, and therefore type.

    Preoccupied by Eurema (Hübn., Lep. 1816).

530. Hewitsonia.

1871. Kirb., Syn. Cat. 426: **Boisduvalii.** Sole species, and therefore type.
  Proposed to replace Corydon, preoccupied.

531. Hexcopteris.*

1816. Hübn., Verz. 22: Endymion (Endymiæna), Cupido (Cupidina).
  This name falls before Helicopis. See also Erotion.

532. Hipio.

1816. Hübn., Verz. 56: **Constantia** (Constantina), Leda.
1865. Herr.-Schaeff., Prodr. i. 61: employs it for other butterflies, Crishna and a MS. species.
1868. Butl., Ent. Monthl. Mag. iv. 194: designates Constantia as the type.

533. Hiposcritia.

1832. Gey. in Hübn., Zutr. iv. 16: **Pandione.** Sole species, and therefore type.

534. Hipparchia.

1807. Fabr., Ill. Mag. vi. 281: Hermione, Statilimus (Fauna), Maera, Ligea, Epiphron, Galathea, Tithonus (Pilosellæ), **Hyperanthus**, Rumina.
1815. Leach, Edinb. Encycl. 717: uses it for Galathea, Hyperanthus, Tithonus (Pilosellæ), and others not of Fabricius' list.
1816. Ochs., Schmett. Eur. iv. 19: divides the group into seven "families," and places in it all the European Satyrids.
1816. Hübn., Verz. 57: uses it for Statilimus (Arachne) and others not used by Fabricius.
1828. Curtis, Brit. Ent., pl. 205: designates Jurtina (Janira) as type, but it was not one of the Fabrician species.
1837. Sodoffsk., Bull. Mosc. x. 81: proposes to replace the name by Melania (q. v.).
1840. Westw., Gen. Syn. 88: specifies Megæra as type.
1844. Doubl., List Br. Mus. 129: uses it for a large number, including Statilimus (Fauna) and others, but not Hyperanthus.
1858. Ramb., Cat. Lép. Andal. 22: uses it for five species, including only Tithonus of those mentioned by Fabricius.
1868. Butl., Ent. Monthl. Mag. iv. 194; Cat. Sat. 50: specifies Hermione (fagi) as type, but incorrectly.

1871. Kirb., Syn. Cat. 80: uses it for Hermione, Statilimus, Hyperanthus, and others.

The Fabrician species belong to as many genera. These were at once restricted by Leach's action in 1815 to three. Hübner therefore misapplied it in 1816, as did Curtis in 1828, and Butler in 1868. Of the species mentioned by Leach, Galathea belongs to Agapetes (1820) and Tithonus to Pyronia (1816), so that Hyperanthus virtually became the type in 1820; consequently the name is also misapplied by Doubleday in 1844 and Rambur in 1858. See Aphantopus and Enodia.

535. HIPPARCHIOIDES.*

1867. Butl., Ann. Mag. Nat. Hist. [3] xix. 125: *Merope*, Philerope, Banksii (Banksia), mirifica.
1868. Ib., Cat. Sat. 99, note: specifies Merope as type.

Falls before Heteronympha (q. v.). See also Tisiphone.

536. HISTORIS.

1816. Hübn., Verz. 35: Odius (Odia), Marthesia.

If Marthesia is congeneric with Ide, which was made type of Sideronc in 1822-26, then Odius must be type of Historis, and Aganisthos must fall. If generically separable, Marthesia must be taken as type of Historis and Odius of Aganisthos. The writer has not the means of determining this point.

537. HOLOCHILA.*

1862. Feld., Verh. zoöl.-bot. Gesellsch. Wien, xii. 490: absimilis, Erinus.
1865. Ib., Reise Novara, 261: absimilis.

The name is proposed for Erina (q. v.), improperly formed; but it is preoccupied in Mammals (Brandt, 1835). See also Polycyma.

538. HOMŒONYMPHA.

1867. Feld., Reise Novara, 487: **pusilla**. Sole species, and therefore type, as stated by Butler (Cat. Sat. etc.).

539. HYADES.

1832. Boisd., Voy. Astrol. 157: Urania (Jairus), **bioculatus** (Indra).
1836. Ib., Spec. gén., pl. 9 B.: Horsfieldii.

Bioculatus may be taken as the type.

540. HYALIRIS.

1870. Boisd., Lép. Guat. 33: **Coeno**. Sole species, and therefore type.

Does this fall before Hypothyris?

541. Hyalites.

1848. Doubl., Gen. Diurn. Lep. i. 140: I. Horta, Quirina (Dice), Ranavalona, Ignati, Hova, Mahela (Neobule, Mahela), Camœna, Andromache, and two MS. species; II. **Lycia.**

Lycia, as the species of the second section which is illustrated by Doubleday, may be taken as the type.

542. Hyantis.

1863. Hewits., Exot. Butt. iii. 68: **Hodeva.** Sole species, and therefore type.

Used in same way by Herrich-Schaeffer and Kirby.

543. Hydrænomia.

1870. Butl., Ent. Monthl. Mag. vii. 99: to supplant Udranomia as more orthographic. Hence **Orcinus** is the type, as designated by Butler for Udranomia.

544. Hylephila.

1820. Billb., Enum. Ins. 81: comma, **Phylæus,** sylvanus, and two MS. species.

Comma became the type of Erynnis in 1832, sylvanus that of Augiades in 1850, so that Phylæus must be taken as the type of this. See Euthymus.

545. Hymenitis.

1816. Hübn., Verz. 8: **diaphanus** (diaphane), Sao.
1844. Doubl., List Br. Mus. 59: uses it for diaphanus (diaphana) and some unnamed species.

It has been used in the same sense by Doubleday (Genera), Bates, and Herrich-Schaeffer (Prodr.). See Godyris.

546. Hypanartia.

1821–25. Hübn., Exot. Schmett. ii.: **Paullus** (Tecmesia), Delius (Demonica), Hippomene.
1827–37. Gey. in Hübn., Exot. Schmett. iii.: Hyppoclus (Hippocla).
1871. Kirb., Syn. Cat. 180: uses it for all of Hübner's species and others.

Paullus may be taken as the type. See Eurema.

547. Hypanis.*

1833. Boisd., Ann. Mus. Hist. Nat. 203: *Ilithyia* (Anvatara). Sole species, and therefore type.
1836. Ib., Spec. gén., pl. 5 B.: Ilithyia (Polinice).

Since used in same sense by Doubleday, Westwood, Kirby, and others; but the name must fall before Byblia (q. v.).

548. HYPATUS.

1825. Hübn., Catal. Franck, 85: Celtis, **Carinenta**.
: Celtis being already type of Libythea, Carinenta must be taken as the type of this. See Hecaerge and Libythea.

549. HYPERMNESTRA.

1851. Heyd., Lep. Eur. Cat. 3d ed. 16: **Helios**.
1852. Westw., Gen. Diurn. Lep. 530: the same.
: Subsequently used similarly by Felder, Kirby, etc. See Ismene.

550. HYPHILARIA.

1816. Hübn., Verz. 26: **Nicias** (Nicia). Sole species, and therefore type.
1867. Bates, Journ. Linn. Soc. Lond. ix. 420: employs it for the same and others, in which he is followed by Kirby (Syn. Cat.).

551. HYPNA.

1816. Hübn., Verz. 56: **Clytemnestra**. Sole species, and therefore type.
: Used in the same sense by Westwood, Felder, Butler, and Kirby. See Hecalene.

552. HYPOCHRYSOPS.

1865. Feld., Reise Novara, 251: Doleschalii, Theon, **Anacletus**, Eucletus, Pythias, Protogenes, Chrysanthis.
1871. Kirb., Syn. Cat. 378: employs it for the same and others.
: Anacletus may be taken as the type.

553. HYPOCYSTA.

1850–51. Westw., Gen. Diurn. Lep., pl. 67: **Euphemia**. Sole species, and therefore type.
1851 (June). Ib., ib. 397: Irius, Euphemia.
1865. Herr.-Schæff., Prodr. i. 60: employs it for this and others, including Irius (Adiante).
1868. Butl., Ent. Monthl. Mag. iv. 196; and Cat. Sat. 167: wrongly specifies Irius as type, for the context shows that the plate was printed before the text.

554. HYPOLIMNAS.

1816. Hübn., Verz. 45 (spelled both Hypolimnas and Hipolimnas): Antilope, Alimena (Velleda, Alimena, Porphyria), Bolina (Eriphile, Perimele, Manilia, Antigone, Alcmene, Iphigenia), **Pandarus** (Pipleis).

1822–25. Ib., Exot. Schmett. ii.: Pandarus (Pipleis).
1871. Kirb., Syn. Cat. 224: employs it for the same and others.
> Most of the species fall into the earlier Apatura, but the name may be retained for Pandarus and its allies, in accordance with Hübner's later use of it.

555. HYPOLYCÆNA.

1862. Feld., Wien. Ent. Monatschr. vi. 293: Tmolus, **Sipylus** (Tharrytas), Astyla.
1871. Kirb., Syn. Cat. 406: employs it for the same and others.
> Sipylus may be taken as the type.

556. HYPOPHYLLA.

1836. Boisd., Spec. gén., pl. 4 C.: **Zeurippe** (Zeurippa).
1847. Doubl., List Br. Mus. 9: employs it for this and others.
1867. Bates, Journ. Linn. Soc. Lond. ix. 446: uses it similarly.

557. HYPOTHYRIS.

1822. Hübn., Index, 5: **Ninonia.** Sole species, and therefore type.
> See Hyaliris.

558. HYREUS.*

1816. Hübn., Verz. 70: Lingeus, Palemon, Misenes.
> The name is preoccupied in Birds. (Steph. 1815).

559. IDAIDES.

1816. Hübn., Verz. 85: **Codrus**, Nireus.
> Codrus may be taken as the type.

560. IDEA.*

1807. Fabr., Ill. Mag. vi. 283: *Idea*. Sole species, and therefore type.
> Subsequently used by Godart and others, but the name cannot stand, from having been borrowed from the species on which it is founded. See Nectaria and Hestia.

561. IDEOPSIS.

1858. Horsf., Cat. Lep. E. Ind. Co. i. 133: **Gaura**, Daos.
1871. Kirby, Syn. Cat. 2: employs it for the same and others.
> Gaura may be taken as the type.

562. IDIOMORPHUS.*

1861. Doum., Rev. Mag. Zoöl. [2] xii. 174: *Hewitsonii*. Sole species, and therefore type, as subsequently specified by Butler.
> Mr. Kirby informs me that the name is preoccupied in Coleoptera (Chaud. 1846). See Bicyclus.

563. IDIONEURA.

1867. Feld., Reise Novara, 474: **Erebioides.** Sole species, and therefore type, as stated by Butler and used by Kirby.

564. IDMAIS.

1836. Boisd., Spec. gén. 584: I. **Chrysonome**, Fausta; II. Phisadia, Amata (Calais), Dynamene.

Subsequently used in same sense by Doubleday, Kirby, and others. The generic name Colotis being applicable to the second section of Boisduval's group, Idmais may be restricted to the first, with Chrysonome for its type.

565. ILERDA.*

1847. Doubl., List Br. Mus. 25: *Epicles* and some MS. species. Epicles is therefore the type.

Used in same sense by Hewitson, Herrich-Schaeffer, and Kirby; but the name falls before Heliophorus (q. v.).

566. ILIADES.

1816. Hübn., Verz. 88: Polymnestor, **Memnon** (Ancaeus, Laomedon, Memnon), Agenor [placed also by Hübner, in same work, in Achillides] (Agenor, Mestor), Protenor.

Memnon may be taken as the type. See Papilio.

567. IMELDA.

1870. Hewits., Equat. Lep. iv. 56: **Glaucosmia.** Sole species, and therefore type.

Used in same sense by Kirby.

568. INACHIS.*

1816. Hübn., Verz. 37: *Io.* Sole species, and therefore type.

Subsequently used by Stephens in the same sense. The name, however, falls before Hamadryas (q. v.), and is also preoccupied through Inachus (Fabr., Crust. 1798).

569. INCISALIA.

1872. Minot in Scudd., Rev. 31: Augustinus (Augustus), **Niphon**, Irus (Irus), Henrici (Irus). Type specified by Scudder as Niphon.

570. IOLAUS.

1816. Hübn., Verz. 81: **Helius** (Eurisus). Sole species, and therefore type, as stated by Hewitson (Ill. Diurn. Lep. 1865, 40).

1847. Doubl., List Br. Mus. 26: employs it for Hymen (Liger) and others, not including Helius.

It is used in Hübner's sense by several authors.

571. Iphias.*

1836. Boisd., Spec. gén. 595: Glaucippe, Leucippe.

Used by Doubleday (1844), but falls before Hebomoia, and the name is preoccupied through Iphius (Dej., Col. 1833).

572. Iphiclides.

1816. Hübn., Verz. 82: Dolicaon, Antiphates, Nomius (Meges), Protesilaus, **Podalirius**, Ajax, Aristeus, Sinon, Antiphates (Pompilius), Antheus, Agamemnon.
1850. Steph., Cat. Brit. Lep. 251: employs it, as does Kirby in 1858, for Podalirius (Podalirius, Feisthamelii), so that this becomes the type, as stated by Scudder (1872).

See Podalirius and Papilio.

573. Isapis.

1847. Doubl., List Br. Mus. 18: **Agyrtus.** Sole species, and therefore type.

Used for same species by Westwood, Bates, and Kirby.

574. Ismene.

1820–21. Swains., Zoöl. Ill. i. 16: **Oedipodia.** Sole species, and therefore type.
1846. Nickerl, Stett. Ent. Zeit. vii. 207: employs it for Helios, a totally different insect. See Hypermnestra.
1848. Ménétr., Mem. Acad. St. Petersb. [6] Sc. Nat. vi. 274: the same.
1852. Westw., Gen. Diurn. Lep. 514: employs it in the Swainsonian sense for a dozen species.
1856. Gray, Pap. Brit. Mus. 77; and List Pap. 92: uses it again for Helios.
1869. Herr.-Schaeff., Prodr. iii. 54: without indicating any species, uses it as Swainson does.
1871. Staud., Catal. 2: again reverts to Nickerl's use.
1871. Kirb., Syn. Cat. 581: uses it correctly.

575. Isodema.

1863. Feld., Wien. Ent. Monatschr. vii. 109, note: proposes the name for Paraplesia (preoccupied), without indication of species. **Adelma**, being the type of Paraplesia (q. v.), becomes the type of this.

576. Isoteinon.

1862. Feld., Wien. Ent. Monatschr. vi. 30: **lamprospilus** (lamprosilus). Sole species, and therefore type.
1871. Kirb., Syn. Cat. 625: the same and another species.

577. Issoria.

1816. Hübn., Verz. 31: Egista, Iole (Anticlia), **Lathonia.**
1850. Steph., Cat. Brit. Lep. 14: uses it for Lathonia only.
> In this he is followed by Kirby (1858), and this therefore becomes the type.

578. Itanus.

1861. Feld., Neues Lep. 34: Aconthea, Garuda, Phemius, Salia, **Anosia.** Anosia may be taken as the type.
> The name is too close to Itamus (Schmidt-Goebel, Col. 1846).

579. Ithobalus.

1816. Hübn., Verz. 88: **Polydamas,** Crassus, Belus, Lycidas (Lycidas, Erymanthus), Numitor.
> Polydamas may be taken as the type.

580. Ithomeis.

1862 (Sept.). Bates, Trans. Linn. Soc. Lond. xxiii. 541: **Aurantiaca,** Stalachtina, Heliconina, mimica, Satellites.
> Aurantiaca may be taken as the type. See Ithomiopsis.

581. Ithomia.

1816. Hübn., Verz. 9: Drymo, Euritea, **Doto,** Cymo.
1821. Ib., Index: Cymo, Doto.
1822? Ib., Samml. Exot. Schmett. text: Doto only, which thereby becomes type. [See Note, p. 293.]
1844. Doubl., List Br. Mus. 57: Drymo, Euritea, and others.
1847. Ib., Gen. Diurn. Lep. 125: uses it in the same sense.
1862. Bates, Linn. Trans. xxiii. 537: Doto, Cymo, and others.
1871. Kirb., Syn. Cat. 26: all of Hübner's and others.
1872. Butl.-Druce, Cist. Ent. i. 95: specify Drymo as type.

582. Ithomiola.

1865. Feld., Reise Novara, 311: **floralis.** Sole species, and therefore type.
> Used for same species only, by Bates and Kirby. See Compsoteria.

583. Ithomiopsis.*

1862 (Dec.). Feld., Wien. Ent. Monatschr. vi. 411: Corena, Astræa.
> Stated by Bates to be synonymous with Ithomeis (q. v.).

584. ITUNA.

1847 (Aug.) Doubl., Gen. Diurn. Lep., pl. 17: **Phenerate**. Sole species, and therefore type.
1847 (Oct.) Ib., ib. i. 113: Lamirus? (Lamyra), Phenerete, Ilione.
    Subsequently used in the same sense.

585. IXIAS.

1816. Hübn., Verz. 95: **Pyrene** (Pyrene, Anexibia, Ænippe), Marianne (Bebryce, Mariane).
1870. Butl., Cist. Ent. i. 37, 48: suggests Pyrene as type.
1871. Kirb., Syn. Cat. 497: employs it for both of Hübner's and others.
    See Thestias.

586. JÆRA.*

1816. Hübn., Verz. 38: Opis, Afer (Afra), Crithea.
1850. Westw., Gen. Diurn. Lep. 269 [Iera]: employs it for Crithea and Cœnobita.
1861. Feld., Neues Lep. 30: using the genus in Westwood's sense, separates the two species into two sections.
1869. Butl., Cat. Fabr. Lep. 102 [Iera]: uses it for the same species.
    The name, however, is preoccupied through Gæra [Scr. Jæra, Agass. Nomencl.] (Leach, Crust. 1815). See Catuna and Euomma.

587. JALMENUS.

1816. Hübn., Verz. 75: **Evagoras**, Venulius.
1847. Doubl., List Br. Mus. 28 [Ialmenus]: employs it for Evagoras and others, excluding Venulius.
    In this sense it has also been used by Herrich-Schaeffer, Hewitson [Ialmenus], and Kirby [Ialmenus], Evagoras therefore is the type. See Austromyrina.

588. JAMIDES.

1816. Hübn., Verz. 71: roboris (Evippus), Ethemon, **Bochus**.
    Bochus may be taken as the type.

589. JASIA.*

1832-33. Swains., Zoöl. Ill. ii. 90: *Jason* (Jasius). Sole species, and therefore type.
    The name, being derived from the species on which the genus is grounded, falls. See Charaxes and Paphia.

590. JASONIADES.

1816. Hübn., Verz. 83: Alexanòr, **Glaucus**, [placed also by Hübner in Euphœades in the same work!] (Turnus), Machaon, Xuthus.

1850. Steph., Cat. Brit. Lep. 2 [Jasonides]: Machaon.
So also Kirby (1858).

Machaon, however, had already been made the type of Princeps: the other species, excepting Glaucus, are congeneric, and hence this must be taken as the type. See Euphœades.

591. JUNONIA.

1816. Hübn., Verz. 34: Aonis, **Lavinia**, Orithya (Orithya, Ocyale, Isocratia), Clelia, Erymanthis (Lotis), Œnone.
1849. Doubl., Gen. Diurn. Lep. 206: places in his typical section all the species of Hübner excepting Erymanthis, besides others not mentioned by him.
1861. Feld., Neues Lep. 13: divides the group into two sections, in the second of which he places two species, referred by Doubleday to his typical section. None of Hübner's species are specially designated.
1871. Kirb., Syn. Cat. 186: follows Doubleday.
1872. Scudd., Syst. Rev. 22: designates Lavinia as the type.
See Alyconeis.

592. KALLIMA.

1849. Doubl., Gen. Diurn. Lep., pl. 52: Rumia, **Paralekta**.
1850. Westw., Gen. Diurn. Lep. 324: nine species are given, including the above.

The question of type is a somewhat peculiar one. The "Paralekta" of Doubleday is considered by Westwood to be distinct from "Paralekta" of Horsfield, and the same as "Horsfieldii" of Kollar. Kirby (Syn. Cat. 193), on the other hand, makes "Paralekta" of Doubleday the same as the "Paralekta" of Horsfield; and the "Paralekta" of Westwood (together with the "Horsfieldii" of Kollar), the same as the "Inachis" of Boisduval, placed as a possible synonyme of "Paralekta." Now Westwood regards *his* "Paralekta" as type. If, however, Westwood's "Paralekta" is not the "Paralekta" of Doubleday, it could not become the type of the genus, because not mentioned in the first instance. If the same, it would have to be taken as type; consequently it would best be considered the type. The question, however, is still further complicated by the following: —

1861. Feld., Neues Lep. 14: refers to it only Inachis and Rumia.

If Inachis (which includes the "Paralekta" of Westwood) is distinct from the Paralekta of Doubleday, and Westwood's decision of a type is thereby ruled out of place, then Rumia becomes the type. The question is, in reality, of little importance, since all the species mentioned are congeneric in the strictest sense, and variety of opinion as to specific alliances does not affect the generic nomenclature.

1871. Kirb., Syn. Cat. 193: employs it for all these and another.

593. KRICOGONIA.

1863. Reak., Proc. Ent. Soc. Phil. ii. 355: **Lyside**, Menippe (Leachiana). Lyside specified as type.
1870. Butl., Cist. Ent. i. 36, 46: Lyside specified as type.

594. LACHNOPTERA.

1847. Doubl., Gen. Diurn. Lep., pl. 22: **Iole**. Sole species, and therefore type.
1848. Ib., ib. 161: the same.

595. LÆOSOPIS.

1858. Ramb., Cat. Lép. Andal. i. 33: **roboris**. Sole species, and therefore type.
1871. Kirb., Syn. Cat. 377: the same, and another.

596. LAERTIAS.

1816. Hübn., Verz. 84: Ulysses (Ulysses, Diomedes), **Philenor**, Menestheus (Cresus), Palinurus (Regulus), Polytes (Pamnon, Cyrus), Merope (Brutus).
1872. Scudd., Syst. Rev. 43: specifies Philenor as the type.

597. LAMPIDES.

1816. Hübn., Verz. 70: Numereus, **Ælianus** (Zethus), Helius, Balliston (Baaliston), Bœticus, Plato, Archias (Archius), Celeno (Celerio), Aratus.
1869. Butl., Cat. Fabr. Lep. 163: employs it for nineteen species, including Minereus, Ælianus, Bœticus, Plato, and Celeno.
1870. Newm., Brit. Butt. 117: employs it for Bœticus only.

It cannot be employed for Bœticus, as this became in 1810 the type of Polyommatus. Ælianus may be taken as the type.

598. LAMPROPTERA.*

1832. Gray, in Griff. An. Kingd., pl. 102, fig. 4: *Curius*. Sole species, and therefore type.

The same species is the type of Leptocircus (q. v.) published at about the same time: perhaps it is impossible to discover which is earlier, but this name is too close, in any case, to Lampropteryx (Steph., Lep. 1829) to stand. Leptocircus is also preferred to this by Gray himself in 1856 (Pap. Brit. Mus.).

599. LAMPROSPILUS.

1832. Gey. in Hübn., Zutr. iv. 80: **Genius**. Sole species, and therefore type.

Subsequently used in same sense by Herrich-Schaeffer and Kirby.

600. LAOGONA.*

1836. Boisd., Spec. gén., pl. 6 B.: *Hypselis.* Sole species, and therefore type.

Subsequently used in same sense by Doubleday and Felder, but the name falls before Symbrenthia (q. v.).

601. LAPARUS.*

1820. Billb., Enum. Ins. 77: Rhea (Sara), Erato (Doris), Phyllis, Melpomene.

The name falls before Sicyonia, Migonitis, and Sunias.

602. LARINOPODA.

1871. Butl., Trans. Ent. Soc. Lond. 172: **lycænoides.** Sole species, and therefore type.

603. LASAIA.*

1867. Bates, Journ. Linn. Soc. Lond. ix. 397: Meris, Cleades (Cleadas).
1871. Kirb., Syn. Cat. 321: the same.

But this name cannot stand, because preoccupied through Lasæa (Brown, Moll. 1827) and Lasia (Wied., Dipt. 1824, and Hope, Col. 1840).

604. LASIOMMATA.

1840. Westw. in Westw.-Humphr., British Butterfl. 65: Ægeria, **Megæra.**
1844. Doubl., List Br. Mus. 134: employs it for Ægeria, Megæra, and other insects not specified by Westwood.
1850. Steph., Cat. Brit. Ent. 6, 254: employs it for Ægeria, Megæra, and Mæra only.
1851. Westw., Gen. Diurn. Lep. 385: employs it for the same and others.

As Ægeria is the type of Pararge, Megæra must be taken as the type of this genus. Butler, in his Catalogue of Satyridæ and elsewhere, has sunk this name under Pararge, apparently on the false principle that the first species must be taken as the type; and he has founded on the second species of this list, and on others, a genus Amecera (q. v.), which must certainly fall, unless some of its other species are generically distinct from Megæra.

605. LASIOPHILA.

1859. Feld., Wien. Ent. Monatschr. iii. 325: **Cirta,** Circe. Felder remarks that the species resemble, in habitus and coloring, the species of Pronophila of the group of Zapatoza.

1867. Butl., Ann. Mag. Nat. Hist. [3] xx. 268; also (1868) Ent. Monthl. Mag. iv. 196, and Cat. Sat. 181: specifies Zapatoza as type, of course erroneously.
1871. Kirb., Syn. Cat. 107: employs the name for all the species mentioned above, and others.
Cirta may be considered the type.

606. Lebadea.

1861. Feld., Neues Lep. 28: **Ismene**, Alankara, Martha.
1871. Kirb., Syn. Cat. 230: the above and Paduka.
Ismene may be taken as the type.

607. Lemonias.

1806. Hübn., Tent. 1: **Maturna**. Sole species, and therefore type.
1818. Ill., Wied. Zoöl. Mag. i. ii. 99: Lamis and others, wholly unrelated to the above.
1847. Doubl., List Br. Mus. 16: uses it in the Illigerian sense for Epulus and others.
1851. Westw., Gen. Diurn. Lep. 457: uses it in a similar way for Chia and six others.
1867. Bates, Journ. Linn. Soc. Lond. ix. 446: extends it greatly, also using it for the Vestales.
1871. Kirb., Syn. Cat. 322: uses it in the Westwoodian sense, and refers the genus to him!
See also Polystichtis, Calospila, Melitæa and Mellida.

608. Leódonta.

1870. Butl., Cist. Ent. i. 34, 40: **Dysoni**, Tagaste, Tellane. Dysoni specified as type.

609. Leonte.*

1816. Hübn., Verz. 52: Menelaus (Nestira), Achilles (Deidamia), Menelaus (Menelae), Achilles (Achilleja), Hecuba, Telemachus (Telemache).
One of the synonymes of Achilles is Leonte Hübn. The generic name being therefore drawn from, or at least the same as, one of the names previously in use for one of the species upon which it is founded, it must be dropped.

610. Lepricornis.

1865. Feld., Reise Novara, 307: **melanchroia**. Sole species, and therefore type.
Used for this species only, by Bates and Kirby.

611. LEPTALIS.

1823. Dalm., Anal. Ent. 40: **Astynome**, Amphione. Astynome specified as type.
1836. Boisd., Spec. gén. 412: uses it for the above and many others.
1844. Doubl., List Br. Mus. 22: makes a similar use of it.
1847. Ib., Gen. Diurn. Lep. 35: uses it similarly.
See Hemerocharis.

612. LEPTIDIA.

1820. Billb., Enum. Ins. 76: sinapis. Sole species, and therefore type.
    Never since used, but should certainly be restored. See Leptoria and Leucophasia.

613. LEPTOCIRCUS.

1832–33. Swains., Zoöl. Ill. ii. 106: **Curius**. Sole species, and therefore type.
    Frequently used since in the same sense. See Lamproptera.

614. LEPTONEURA.*

1857. Wallengr., Rhop. Caffr. 31: *Clytus*. Sole species, and therefore type, as stated by Butler.
    It should fall, however, before Dira (q.v.).

615. LEPTOPHOBIA.

1870. Butl., Cist. Ent. i. 35, 45: **Eleone**, Penthica (Pentica), Balidia, Aripa (Arapa), Pylotis. Eleone specified as type.

616. LEPTOPTERA.*

1842. Boisd. in Lucas, Ann. Soc. Ent. Fr. [1] xi. 298: *decora*. Sole species and designated type.
    The species was at that time inedited, and before it was published Boisduval had changed the name to Amnosia (q.v.).

617. LEPTORIA.*

1841. Westw., Brit. Butt. 31: *sinapis* (candida). Sole species, and therefore type.
    Falls before Leptidia. See also Leucophasia and Leptosia, for the latter of which it was probably a misprint.

618. LEPTOSIA.

1816. Hübn., Verz. 95: sinapis (lathyri), Alcesta, **Xiphia** (chlorographa), Brephos.
1858. Kirb., List Brit. Rhop.: employs it for sinapis (candida, erysimi).

1870. Butl., Cist. Ent. i. 39, 54: specifies sinapis (lathyri) as type, but wrongly, as this was already the type of three different genera! See Leptidia.

Sinapis was taken as type of Leptidia in 1820, Brephos has belonged to Leucidia since 1847, Alcesta and Xiphia are congeneric, and Xiphia may be taken as the type. See Nina and Nychitona.

619. LEREMA.

1872. Scudd., Syst. Rev. 61: **Accius**, Ilianna, Pattenii. Accius specified as type.

620. LERODEA.

1872. Scudd., Syst. Rev. 59: **Eufala**, fusca, Inca. Eufala specified as type.

621. LETHE.

1816. Hübn., Verz. 56: **Europa**. Sole species, and therefore type, as stated by Butler. See Debis.

622. LETHITES (fossil). See Satyrites.

623. LEUCIDIA.

1847. Boisd. in Doubl., Gen. Diurn. Lep. 77: **Elvina**, Brephos.
1867. Herr.-Schaeff., Prodr. ii. 8: Brephos, Leucoma (Elphos).
1870. Butl., Cist. Ent. 35, 43: specifies Leucoma (Elphos) as type, but of course erroneously.

Elvina may be taken as the type.

624. LEUCOCHITONEA.

1857. Wallengr., Rhop. Caffr. 52: **Levubu**. Sole species, and therefore type, as stated by Butler.

Since used by authors in too extended a sense.

625. LEUCONEA.*

1837. Donz., Ann. Soc. Ent. Fr. vi. 80: *cratægi*. Sole species, and therefore type.
1858. Ramb., Catal. Lép. Andal. 54: uses it in the same way.

The name falls before Aporia.

626. LEUCOPHASIA.*

1827. Steph., Ill. Brit. Ent. Haust. i. 24: *sinapis*. Sole species, and therefore type, as specified by Westwood (Gen. Syn. 87).

Used in same sense by many subsequent authors. Falls before Leptidia. See also Leptoria.

627. LEUCOSCIRTES.*

1872. Scudd., Syst. Rev. 52: *ericetorum*, Arsalte (nivea), Oceanus. Ericetorum specified as type.
   The name falls before Heliopetes.

628. LEUCOTHYRIS.*

1870. Boisd., Lép. Guat. 32: *Ilerdina*. Sole species, and therefore type.
   This name is too close to Leucothyreus (MacL., Col. 1819) to be used.

629. LEXIAS.

1832. Boisd., Voy. Astrol. 125: **Æropus**. Sole species, and therefore type.
1861. Feld., Neues Lep. 36: places this species in a first section, Dirtea (Dirtea, Boisduvalii) in a second.

630. LIBYTHEA.

1807. Fabr., Ill. Mag. vi. 284: **celtis**, Carinenta.
1810. Latr., Consid. 440: specifies celtis as the type.
1820. Billb., Enum. Ins. 79: changes the name, for no reason, to Chilea.
1828. Boit., Man. Ent. ii. 299 [Libythæus]: celtis.
   It has been used constantly by all authors in much the same sense.
1872. Crotch, Cist. Ent. i. 66: states that celtis is the type, through Latreille, 1810.
1872. Scudd., Syst. Rev. 28: specifies Carinenta as type, erroneously. See Hypatus and Hecaerge.

631. LIBYTHINA.

1861. Feld., Neues Lep. 49: **Cuvieri**. Sole species, and therefore type.

632. LICINIA.*

1820–21. Swains., Zoöl. Ill. i. i. 15: *Melite*. Sole species and designated type.
   Subsequently, in the same series (i. ii. 91; i. iii. 124), Amphione and Critomedia (Crisia) are given. An allied species is Licinia of Cramer, doubtless intended by Swainson to be included in the group, and from which the name was drawn; on which account the name should be dropped. It is also preoccupied in Mollusks (Brown, 1756). See Enantia.

633. LIMENITIS.

1807. Fabr., Ill. Mag. vi. 281: populi, Niavius, **Camilla**.

1815. Leach, Edinb. Encycl. 718: employs it for Camilla only, so that this becomes the type. [See Note, p. 293.]
1816. Dalm., Vetensk. Acad. Handl. xxxvii. 56 [Limonitis]: specifies populi as the type. See Najas.
1816. Hübn., Verz. 44: employs it for Camilla, populi, and two others.
1820. Billb., Enum. Ins. 78 [Limonetes]: uses it for populi and others.
1832. Dup., Pap. France, Diurn. Suppl. 400: uses it for Sibylla, Camilla, Lucilla, and aceris.
1832. Renn., Consp. 11 [Leminitis]: populi, etc.
1840. Westw., Gen. Syn. 87: specifies Camilla as type.
1844. Doubl., List Br. Mus. 93: employs it for Camilla and others, while populi is placed under Nymphalis.
1850. Westw., Gen. Diurn. Lep. 274: regards populi as the type.
1872. Crotch, Cist. Ent. i. 66: regards populi as the type, through Dalman, 1816, overlooking Leach's previous action.

634. LIMNÆCIA.*

1872. Scudd., Syst. Rev. 26: *Harrisii.* Sole species and designated type.

This falls before Cinclidia (q.v.).

635. LIMNAS.

1806. Hübn., Tent. i.: **Chrysippus.** Sole species, and therefore type.
1836. Boisd., Spec. gén., pl. 4 C.: Pixe, a totally different insect from that of Hübner. See Melanis.
1840. Blanch., An. Art, iii. 464 [Lynmas]: Jarbus (Electron). Closely allied to Boisduval's species.

Doubleday, Westwood, Bates, Herrich-Schaeffer, Felder, and Kirby, have all since used it in the Boisduvalian sense. But as Chrysippus is generically distinct from Plexippus, Limnas will stand for the former.

636. LINCOYA.

1871. Kirb., Syn. Cat., App. 649: **Pharsalia,** Felderi.
1873. Ib., Zoöl. Rec. for 1871, 360: specifies Pharsalia as type.

Correctly, as this was the type of Antigonis (q.v.), which Lincoya was intended to supplant.

637. LIMOCHORES.

1872. Scudd., Syst. Rev. 59: Mystic, bimacula, **Manataaqua,** Taumas, Arpa, Pilalka (Palatka), and a MS. species. Manataaqua is specified as type.

638. Liphyra.

1864. Westw., Proc. Ent. Soc. Lond. xxxi.: **Brassolis**. Sole species, and therefore type.
   See Sterosis.

639. Liptena.*

1852? Westw., Gen. Diurn. Lep., pl. 77: Abraxas, Acræa.
1852. Ib., ib. 503: used as a synonyme of Pentila.
1865. Hewits., Exot. Butt. iii. 119: employs it for Acræa and others, so that Acræa becomes the type.
1868. Herr.-Schaeff., Prodr. iii. 13: follows Hewitson.
1871. Kirb., Syn. Cat. 335: follows Hewitson.
   The name falls before Pentila and Tingra.

640. Loxura.*

1829. Horsf., Descr. Cat. Lep. E. Ind. Co. 119: *Atymnus*, Pita. Atymnus specified as type.
   Since used in same sense by Boisduval, Duncan, and Westwood, but the name must fall before Myrina (q.v.).

641. Lucia.

1832–33. Swains., Zoöl. Ill. ii. 135: **Aurifer** (Limbaria). Sole species, and therefore type.
   Since used by authors in the same sense.

642. Lucilla.

1870. Hewits., Equat. Lep. iv. 55: **Camissa**. Sole species, and therefore type.

643. Lucinia.

1822–26. Hübn., Exot. Schmett. ii.: **Sida**. Sole species, and therefore type.
   Since used by Westwood, Felder, and Kirby, in same sense. See Autodea.

644. Lycæides.*

1816. Hübn., Verz. 69: *Argyrognomon* (Argus), Argus (Ægon), Optilete (Optilete, Cyparissus).
1850. Steph., Cat. Brit. Lep. 20, 261: employs it for Argus and other species not in Hübner's list. Argus therefore becomes the type.
1872. Scudd., Syst. Rev. 33: specifies Argus as the type.
   The name falls before Rusticus. See also Scolitantides.

645. LYCÆNA.

1807. Fabr., Ill. Mag. vi. 285: I. Acis (Mars), Echion; II. Argiades, (Amyntas), rubi; III. Endymion (Meleager), Arion, Corydon, Thetis (Adonis), Leda (Ledi), virgaureæ, Phlæas.
1815. Oken, Lehrb. i. 717: restricts it mainly to the blues, referring to it all of the species indicated by Fabricius, excepting rubi and the coppers, virgaureæ and Phlæas, and adding others.
1816. Hübn., Verz. 23: employs it for Echerius (Xenodice), which has nothing to do with Fabricius's species.
1824. Curtis, Brit. Ent., pl. 12: designates Phlæas as type, but that is ruled out by Oken's action.
1828. Horsf., Descr. Cat. Lep. E. Ind. Co. 68: restricts it also to the coppers, but, for the same reason, erroneously.
1828. Steph., Ill. Brit. Ent. Haust. i. 79: does the same.
1832. Renn., Consp. 16: the same.
1832. Dup., Pap. France, Diurn. Suppl. 390: uses it for Bœticus and Telicanus, which belong elsewhere. See Polyommatus.
1832–33. Swains., Zoöl. Ill. 132: also designates Phlæas as the type.
1833. Boisd., Nouv. Ann. Mus. Hist. Nat. ii. 171: uses it for Bœticus, Telicanus, and others.
1836. Ib., Spec. gén., pl. 3 B.: gives a figure of Bœtica.
1837. Sodoffsk., Bull. Mosc. x. 81, 96: proposes to change the name to Lycia or Migonitis, preferably the latter.
1839. Ramb., Faune Ent. Andal. 262: restricts it again to the coppers, erroneously.
1840. Westw., Gen. Syn. 88: specifies Phlæas as type.
1847. Doubl., List Br. Mus. 40: employs it for a great number of species, including, of Fabricius's list, Argiades (Amyntas), Endymion (Meleager), Arion, Corydon, Thetis (Adonis).
1852. Westw., Gen. Diurn. Lep. 488: makes a similar but more extended use of it, in which he has been followed by most recent writers.
1871. Kirb., Syn. Cat. 340: restricts it again to the coppers.
1872. Scudd., Syst. Rev. 36: again specifies Phlæas as type.

No restriction of this group within the blues having been effected, the genus may be confined to Endymion and Corydon of the species mentioned by Fabricius, with Endymion for type. See Heodes.

646. LYCÆNESTHES.

1865. Moore, Proc. Zoöl. Soc. Lond. 773: **bengalensis**. Sole species, and therefore type.

647. LYCÆNOPSIS.

1865. Feld., Reise Novara, 257: **Ananga**. Sole species, and therefore type.

648. LYCNNUCUS.

1825. Hübn., Zutr. iii. 24: **Olenus**. Sole species, and therefore type.

649. LYCIA.*

1837. Sodoffsk., Bull. Mosc. x. 81: proposes this name to supplant Lycæna, for etymological reasons.

But these are insufficient, and Lycia is preoccupied in Lepidoptera (Hübn. 1816).

650. LYCOREA.

1847 (July). Doubl., Gen. Diurn. Lep., pl. 16: **Cleobæa** (Atergatis). Sole species, and therefore type.

1847 (Aug.). Ib., ib. 105: Pasinuntia, Ceres, Halia, Cleobœa (Atergatis, Cleobæa).

This name is very close to Lycoris (Sav., Worms, 1817).

651. LYCUS.*

1816. Hübn., Verz. 74: Niphon, *rubi*, Damon (Gryneus).
1850. Steph., Cat. Brit. Lep. 17: employs it for rubi only, which thereby becomes the type.

But the name is preoccupied in Coleoptera (Fabr. 1787).

652. LYMANOPODA.

1851 (Jan.?) Westw., Gen. Diurn. Lep., pl. 67: **Samius**.
1851 (July). Westw., Gen. Diurn. Lep. 401: Samius, Ionius, obsoleta.
1865. Herr.-Schaeff., Prodr. i. 56: employs it for Samius and others not mentioned by Westwood.
1868. Butl., Ent. Monthl. Mag. iv. 196; and Cat. Sat. 168: designates Samius as the type.

See Sarromia.

653. LYROPTERYX.

1851. Westw., Gen. Diurn. Lep. 433: **Apollonia**, Terpsichore.

Apollonia may be taken as the type.

654. MANCIPIUM.

1806. Hübn., Tent. 1: **brassicæ**. Sole species, and therefore type.
1827. Steph., Ill. Brit. Ent. Haust. i. 22: employs it for Daplidice and cardamines, belonging to the same subfamily as Hübner's species.

1829. Horsf., Descr. Cat. Lep. E. Ind. Co. 141 : uses it as a subdivision of Pontia, assigning to it three species which have intimate connection with the above.
1840. Westw., Gen. Syn. 87 : specifies cardamines as type.
1852. Renn., Consp. 4 : follows Stephens.
1850. Steph., Cat. Brit. Lep. 5 : restricts it still further to Daplidice.

See also Ganoris, Pontia, and Pieris.

655. MANIOLA.

1801. Schrank, Fauna Boica, ii. i. 152, 170 : I. Galathea (Galatœa), Ægeria, Megæra, Mæra, Dejanira, Medea, Ligea, Medusa, Jurtina (Lemur), Epiphron (Egea), Manto (Baucis), Hyperanthus, Arcania (Arcanius), Hero, Typhon (Tiphon), Pamphilus, Iphis (Manto), Semele, Phædra, Briseis (Briseis, Janthe), Hermione, Circe (Proserpina); II. Iris (Iris, Jole), Ilia (Julia, Ilia, Clytie).
1815. Oken, Lehrb. i. 732 : employs it for Iris and Ilia.
1816. Hübn., Verz. 64 : uses it for Afer (Phegea) and Phryne, both Satyrids, but neither of them specified by Schrank.

The former, however, is congeneric with those of Schrank's species, for which the name Erebia must be used by the earlier action of Dalman. Hence Hübner's action has no effect upon Schrank's genus.

1829. Meig., Eur. Schmett. i. 104 : employs it for Briseis and very many others.

All his Satyrids are included, excepting those placed by him in Melanargia (Agapetes): comprising, among others, Jurtina, which may be taken as the type. Excluding the second section of Schrank's genus, which belongs to Potamis, there are no less than ten genera represented by the species enumerated in his list. Of these genera, one (Agapetes) was taken out by Billberg, in 1820; one (Erebia) by Dalman, in 1816; and one (Hipparchia), through the action of various writers, in 1820. Most of the others are taken up by Hübner's generic names, so that the choice finally lies between the present group represented by Jurtina, and that for which we have restricted Nytha (q. v.).

1859. Hein., Schmett. Deutschl. u. Schweiz, i. 26 : Dejanira.

But this has belonged to Pararge from its foundation.

1871. Kirb., Syn. Cat. 57 : considers it the same as Erebia.

656. MARICA.*

1816. Hübn., Verz. 63 : Stygne [also given by Hübner in the same work, in his genus Phorcis], Nelo.

The name falls before Erebia. See also Gorgo, Syngea, Phorcis, Epigea, and Orcina.

657. MARIUS.*

1832–33. Swains., Zoöl. Ill. ii. 45: *Chiron* (Cinna). Sole species, and therefore type.
1832–33. Ib., ib. ii. 59: Peleus (Thetys).

As the work was published in parts, Chiron was published before Peleus. Marius is one of the synonymes of Chiron; the generic name being based upon it falls. See also Euglyphus and Megalura.

658. MARMESSUS.

1816. Hübn., Verz. 81: Silenus (Alcides, Corax), Atymnus, **Lisias**.

Silenus and Atymnus belonging to the earlier Myrina, Lisias must be taken as the type of Marmessus.

659. MARPESIA.

1816. Hübn., Verz. 47: Thyonneus (Thyonnea), **Eleucha** (Eleuchea), Iole (Zosteria), Chiron (Chironias), Orsilochus (Cinna).
1844. Doubl., List Br. Mus. 86: employs it for Eleucha and Peleus (Thetis). Eleucha thereby becomes the type.
1850. Westw., Gen. Diurn. Lep. 263: uses it in the same way.

660. MECHANITIS.

1807. Fabr., Ill. Mag. vi. 284: Calliope, **Polymnia**, Erato (Doris), Psidii, Phyllis.
1866. Hübn., Verz. 11: uses it for Eucrate and Polymnia (Lysimnia, Polymnia). Polymnia therefore becomes the type.
1844. Doubl., List Br. Mus. 55: employs it for Polymnia (Lysimnia), and others.
1847. Ib., Gen. Diurn. Lep. 128: divides the eighteen species which he refers to this genus into two sections, placing Polymnia in the first.
1862. Bates, Linn. Trans. xxiii. 528: restricts the group to Doubleday's first section, dividing that again into two sections, of which Mechanitis proper is made to include "Polymnia and its allies."

See also Nereis.

661. MEGALURA.*

1840. Blanch., Hist. Ins. iii. 446: *Coresia*. Sole species, and therefore type.
1871. Kirb., Cat. 220: Coresia and many others.

The name is preoccupied through Megalurus (Vig.-Horsf., Birds, 1820; Agass., Fishes, 1833). See also Euglyphus and Marius.

662. MEGAMEDE.

1816. Hübn., Verz. 50: **Rhetenor** (Rhetenoris, Chalciope). Sole species, and therefore type.

663. MEGASTES.*

1851. Boisd. in Westw., Gen. Diurn. Lep. 346: given as a MS. synonyme of Dynastor (q. v.) by Westwood.

The species of Dynastor were Napoleon and Darius. Megastes, however, was only applied to Napoleon (Napoleo), and hence the name must fall before Dynastor.

1870. Boisd., Lép. Guat. 53: Macrosiris, Darius.

664. MEGATHYMUS.*

1872. Scudd., Syst. Rev. 62: *yuccæ*. Sole species and designated type.

It is not a butterfly.

665. MEGISTANIS.

1844. [Boisd. in] Doubl., List Br. Mus. 109: Acheronta (Cadmus), **Bæotus** (Beotus).
1849. Boisd. in Doubl., Gen. Diurn. Lep., pl. 48: Bœotus (Beotus).
1850. Ib. in Westw., Gen. Diurn. Lep. 311: Bæotus (Beotus), Acheronta (Cadmus), and another.

By the publication of the plates of Doubleday and Westwood's Genera, Bæotus became the type, and in this sense it has been employed by Felder and Kirby. But Boisduval in 1870 (Lép. Guat.) refers Acheronta again to it. Kirby in his Synonymical Catalogue refers the genus to Westwood.

666. MEGISTO.

1816. Hübn., Verz. 54: Euritus (Cymelia), Argante, Canthus (Euridice), **Acmenis**.
1868. Butl., Cat. Sat. 14: specifies Eurytus as the type.
1872. Scudd., Syst. Rev. 6: does the same. But Eurytus is strictly congeneric with Penelope, the type of Cissia, and therefore Acmenis may be taken as the type of Megisto.

667. MEGONOSTOMA.*

1863. Reak., Proc. Ent. Soc. Phil. ii. 356: Cesonia (Cœsonia), Eurydice, Philippa, Helena.
1870. Butl., Cist. Ent. i. 36, 46: specifies Cesonia as the type.
1872. Kirb., Syn. Cat. 489 [Meganostoma]: Cesonia and allies.

The name must fall before Zerene, which, by the foundation of Eurymus, became restricted to this group.

668. MELAMPIAS.

1816. Hübn., Verz. 63: **Hyperbius** (Hyperbia), Mnestra, Epiphron (Rhodia, Janthe), Pharte, Arete.
1850. Steph., Cat. Brit. Lep. 9, 255: uses it for Epiphron (Cassiope), Melampus, and Mnestra.
1858. Kirb., List Brit. Rhop.: does the same.

> But it cannot be used for these and their allies, as they are already taken up by Erebia, and consequently Hyperbius must be taken as the type. See Pseudonympha.

669. MELANARGIA.*

1829. Meig., Eur. Schmett. i. 97: Galathea (Leucomelas, Galathea, Procida, Electra, Galene), Lachesis, Russiæ (Clotho, Japygia), Arge (Simula), Thetis, Occitanica (Syllius).
1861. Staud., Cat. Lep. Eur. 9 [Melanagria]: refers the same and others to it.
1865. Herr.-Schaeff., Prodr. i. 58 [Melanagria]: the same.

> A strictly homogeneous group, so that the name must fall before the earlier Agapetes. See also Arge.

670. MELANIA.*

1837. Sodoffsk., Bull. Mosc. x. 81: proposes to supplant Hipparchia by this word, but without sufficient reason; moreover, it is preoccupied in Mollusks (Sow. 1819).

671. MELANIS.

1816. Hübn., Verz. 25: **Melander** (Melandra), Phereclus (Pherecla), Agyrtus (Agyrte).

> Melander may be taken as the type. See Limnas.

672. MELANITIS.

1807. Fabr., Ill. Mag. vi. 282: Leda, **undularis**.
1809. Latr., Gen. Crust. et Ins. iv. 197: Ariadne (Ariadne, Merione), undularis. The latter therefore is the type.
1828. Horsf., Cat. Lep. E. Ind. Co., expl. pl. 8: undularis only.
1833. Boisd., Ann. Mus. Hist. Nat. ii. 205: uses it for undularis and others.
1844. Doubl., List Br. Mus. 143: makes a similar use of it.
1851. Westw., Gen. Diurn. Lep. 403: uses it in the same manner.
1868. Butl., Ent. Monthl. Mag. iv. 194; Cat. Sat. 1: specifies Leda as the type, doubtless for the single reason that it is the first species mentioned by Fabricius, yet, as seen by the foregoing, erroneously.

1871. Kirb., Syn. Cat. 43: follows Butler.
1872. Ib., Trans. Ent. Soc. Lond. 1872, 115: specifies undularis as type.

673. MELANOCYMA.
1857. Westw., Trans. Ent. Soc. Lond. [N. S.] iv. 186: **Faunula**. Sole species, and therefore type.

674. MELETE.
1832–33. Swains., Zoöl. Ill. 79: **Lycimnia** (Limnobia). Sole species and designated type.
See Daptonoura.

675. MELINÆA.
1816. Hübn., Verz. 11: **Egina**, Clara, Equicola, Euniæ, Irene.
1837. Sodoffsk., Bull. Mosc. x. 80: ignorant of Hübner's use of it, proposes this name to supplant Melitæa.
1844. Doubl., List Br. Mus. 56: employs it for Egina and Clara of Hübner's species, besides others.
It has since been used in the same sense, and Egina may be taken as the type.

676. MELITÆA.*
1807. Fabr., Ill. Mag. vi. 284: Lucina, Didyma (Cinxia), Cynthia, Maturna.
1816. Dalm., Vetensk. Acad. Handl. xxxvii. 57: specifies Leucippe (Athalia) as type, but of course erroneously.
1832. Curt., Brit. Ent. pl. 386: designates Euphrosyne as type.
1837. Sodoffsk., Bull. Mosc. x. 80: proposes to change the name to Melinæa (q. v.).
1840. Westw., Gen. Syn. 88: specifies Cinxia as type.
1872. Crotch, Cist. Ent. i. 66: says that Leucippe (Athalia) is type, through Dalman.
The name, however, falls, because preoccupied through Melitea (Pér.-Les., Acal. 1809). See Lemonias.

677. MELLICTA.*
1820. Billb., Enum. Ins. 77: Maturna, Aurinia (Artemis), Cinxia, Didyma, Dictynna, Athalia, Parthenie, Lucina, and some MS. species.
This name falls before the earlier Lemonias, Schoenis, and Cinclidia.

678. MEMPHIS.
1816. Hübn., Verz. 48: **Polycarmes** (Odilia), Basilia.
Polycarmes may be taken as type.

679. MENELAIDES.

1816. Hübn., Verz. 84: Hector, **Polytes** (Romulus), Demetrius, Theseus, Aristolochiæ (Polydorus), Polytes, Alphenor, Ascanius, Agavus.

Polytes may be taken as the type.

680. MENERIS.

1844. [Boisd. in] Doubl., List Br. Mus. 106: **Tulbaghia.** Sole species, and therefore type.
1849. Boisd. in Doubl., Gen. Diurn. Lep., pl. 46: the same.
1850. Boisd. in Westw., ib. 296: the same.

It has since been used in the same sense. See Æropetes.

681. MESAPIA.

1856. Gray, List Lep. Brit. Mus. 92: **Peloria.** Sole species, and therefore type.
1872. Kirb., Syn. Cat. 510: the same.

The name is very near to Mesapus (Raf., Crust. 1814).

682. MESENE.

1847. Doubl., List Br. Mus. 7: **Phareus**, Thelephus (Telephus), and MS. species.
1851. Westw., Gen. Diurn. Lep. 441: Phareus (Pharea), Thelephus (Telephus), and others.
1867. Bates, Journ. Linn. Soc. Lond. ix. 439: employs it for Doubleday's species and many others.
1871. Kirb., Syn. Cat. 315: Doubleday's species and others.

Phareus may be taken as the type. See Hübner's use of Emesis.

683. MESOPHTHALMA.

1851. Westw., Gen. Diurn. Lep. 455: **Idotea** (Idotæa). Sole species, and therefore type.

684. MESOSEMIA.

1816. Hübn., Verz. 21: Philemon (Icare), Hyphæa (Hiphia), **Philocles** (Philoclessa), Cœa, Ulrica (Ultio), Osinia, Eumene, Crœsus (Capanea), Ephyne, Thymetus (Thymete), Rosina.
1847. Doubl., List Br. Mus. 12: employs it for Philemon, Philocles, Crœsus (Capanea), and a number of unpublished species.
1851. Westw., Gen. Diurn. Lep. 453: uses it for Philocles, and other species not given by both Hübner and Doubleday. This, therefore, becomes the type.

1867. Bates, Journ. Linn. Soc. Lond. ix. 416: employs it for many species, including Philocles.

1871. Kirb., Syn. Cat. 288: follows Bates.

685. MESOTÆNIA.

1871. Kirb., Syn. Cat. 209: **Doris**. Sole species, and therefore type.

Employed in place of Callitænia, preoccupied; but it is itself very close to Mesotena (Eschsch., Col. 1831).

686. MESSARAS.*

1848. Doubl., Gen. Diurn. Lep. 163: *Erymanthis*, Alcippe.

Subsequently used by Felder and Kirby, the latter for Erymanthis only and its allies. But this name must fall, whichever species is chosen as type. See Atella and Cupha.

687. MESTRA.

1822-26. Hübn., Exot. Schmett. ii.: **Hypermnestra** (Hypermestra). Sole species, and therefore type. See also Cystineura.

688. METACHARIS.

1867. Butl., Ent. Monthl. Mag. iii. 174: **Ptolomæus**, Agrius, Cadmeis, regalis, Lucius (Batesii). The first three specified as types.

1868. Bates, Journ. Linn. Soc. Lond. ix. 444: places nine species here, including Ptolomæus and Agrius and other of Bates's species, but excluding Cadmeis, which is placed under Charis.

1871. Kirb., Syn. Cat. 320: uses it in the same way as Bates.

Ptolomæus may be taken as the type.

689. METAMORPHA.

1816. Hübn., Verz. 43: **Sulpitia** (Elissa), Steneles (Sthenele), Dido.

Dido belongs to Colænis, and Sulpitia may be taken as the type of this group, since it is generically distinct from Steneles, the type of Victorina.

690. METAPHELES.

1866. Bates, Ent. Monthl. Mag. iii. 155: **Dinora**. Sole species, and therefore type.

Used for same species by Bates and Kirby.

691. METAPORIA.

1870. Butl., Cist. Ent. i. 38, 51: **Agathon**. Sole species and designated type.

692. METHONA.*

1847 (Oct.). Doubl., Gen. Diurn. Lep. 115: *Themisto.* Sole species, and therefore type.

> Subsequently used by Bates and Herrich-Schaeffer in same sense; but the name must fall before Thyridia, limited at the same time to same group by Doubleday himself. Doubleday also, in the same year, established a genus Methone for an entirely different insect.

693. METHONE.*

1847. Doubl., List Br. Mus. 4: *Cecilia* (Cæcilia). Sole species, and therefore type.
1851. Westw., Gen. Diurn. Lep. 422: the same. See also Table vi. and 533.

> Westwood changes this name to Methonella (q. v.) because preoccupied by Methona, q. v. (Doubl., Lep. 1847), but both bear the same date. If Methona was first published, of course this falls, and Bates and Kirby assume this.

694. METHONELLA.

1852. Westw., Gen. Diurn. Lep. Table vi. and 533: **Cecilia.** Sole species, and therefore type.

> Subsequently used in same way by Bates and Kirby. See Methone.

695. METURA.

1873. Butl., Lep. Exot. 155: Rurina, irrigata, intermedia, Virgo, **Cipris** (bracheolata, Neocypris).

> Cipris may be taken as the type. Is the name too near Mitoura?

696. MICROTIA.

1864–65. Bates, Ent. Monthl. Mag. i. 83: **Elva.** Sole species, and therefore type.

> This name is very close to Microtus (Schrank, Mam. 1789).

697. MIDEA.*

1867. Herr.-Schaeff., Prodr. ii. 16: *Genutia.* Sole species, and therefore type.

> Used also by Kirby; but the name is founded on one of the synonymes of Genutia, and therefore falls. See Anthocharis.

698. MIGONITIS.

1816. Hübn., Verz. 12: Thales, Aœde, **Erato** (Erato, Crenis), Burneyi, Thelxiope, Melpomene (Andremone, Ulrica, Erythræa), Egeria (Isœa).

1837. Sodoffsk., Bull. Mosc. x. 82 : probably ignorant of Hübner's use of this word, proposes to substitute it for Lycæna.

Erato may be taken as the type. See also Crenis and Laparus.

699. MILETUS.

1816. Hübn., Verz. 71 : **Polycletus** (Epopus, Polycletus), Symethus.
1852. Westw., Gen. Diurn. Lep. 502 : employs it for Symethus and three others.
1857. Horsf.-Moore, Cat. Lep. E. Ind. Co. : make a similar use of it.
1871. Kirb., Syn. Cat. 337 : the same.

Symethus would therefore be type; but Boisduval had already selected this as type of Gerydus (q. v.), and therefore Polycletus must be type. See also Symetha.

700. MIMACRÆA.

1872. Butl., Lep. Exot. i. 104 : **Darwinia.** Sole species, and therefore type.

701. MIMONIADES.

1823. Hübn., Zutr. ii. 27 : **Iphinous** (Ocyalus). Sole species, and therefore type.

702. MINETRA.*

1832. Boisd., Voy. Astrol. 126 : Nodrica, *sylvia*.
1844. Doubl., List Br. Mus. 86 : uses it for sylvia, Gambrisius. Sylvia therefore becomes type.

Since used for all these species by Westwood and Felder. The name falls before Parthenos (q. v.).

703. MINOIS.

1816. Hübn., Verz. 57 : **Phædra,** Alcyone, Hermione, Circe (Proserpina), Persephone (Anthe), Briseis, Merope (Œnomais.)
1850. Steph., Cat. Brit. Lep. 254 : employs it for Briseis, Phædra, and Hermione.
1858. Kirb., List Brit. Rhop. : uses it for Phædra only, which therefore becomes type.
1867. Butl., Ent. Monthl. Mag. iii. 279 : employs it for Phædra (Dryas) and others.
1868. Ib., ib. iv. 194 ; and Cat. Sat. 61 : designates Phædra (Dryas) as type.
1872. Scudd., Syst. Rev. 5 : does the same.

704. MITHRAS.

1816. Hübn., Verz. 79: **Nautes** (Nautus), Elis, Meton (Metus), Apidanus (Apidanus, Dorimund).

1869. Butl., Cat. Fabr. Lep. 195: employs it for Pholeus and others, none of which are mentioned by Hübner, although allied to all but the last.

In accordance with Butler's usage, Nautes may be taken as the type. See Molus.

705. MITOCERUS.*

1820. Billb., Enum. Ins. 79: *Phidippus.* Sole species, and therefore type.

The name falls before Amathusia.

706. MITOURA.

1872. Scudd. Syst. Rev. 31: **Damon** (smilacis). Sole species, and therefore type.

1874. Rye, Zoöl. Rec. for 1872, 350: suggests spelling it Mitura.

It is derived from μίτος and οὐρά.

707. MŒRA.*

1816. Hübn., Verz. 51: Aurelius (Aurelia), Phidippus (Phidippe), Adonis (Adonidis) Tullia, Celinde, Automedon (Automedæna).

The name is preoccupied in Crustacea (Leach, 1815).

708. MOLUS.

1816. Hübn., Verz. 78: **Phalanthus** (Phalantus), Ismarus.

Philanthus may be taken as the type. Will it fall before Mithras?

709. MONETHE.

1851. Westw., Gen. Diurn. Lep. 461: **Alphonsus.** Sole species, and therefore type.

Used in same sense by Bates and Kirby.

710. MORITZIA.*

1861. Feld., Wien. Ent. Monatschr. v. 100: *noctula* (paradoxa). Sole species, and therefore type.

The name falls before Hades.

711. MORPHEIS.*

1827–37. Gey. in Hübn., Exot. Schmett. iii: **Ehrenbergii**. Sole species, and therefore type.

Used for same species only, by Doubleday, Westwood, and Felder; but the name is preoccupied in Lepidoptera (Hübn. 1816). See Anemeca.

712. MORPHO.

1807. Fabr., Ill. Mag. vi. 280: **Achilles**, Menelaus, Hecuba.
1815. Oken, Lehrb. i. 733: employs it for Sibylla, Camilla, and populi!
1816. Hübn., Verz. 49: employs it for species of Prepona only.
1820. Oken, Lehrb. f. Schulen, 791: the Fabrician species and others.
1836. Boisd., Spec. gén., pl. 8 B.: employs it for Cytheris, a species allied to the Fabrician.
1844. Doubl., List Br. Mus. 115: uses it for all the species of Fabricius and others.
1851. Westw., Gen. Diurn. Lep. 337: the same. On p. 341 Achilles is specified as the type.
1872. Crotch, Cist. Ent. i. 65: specifies Achilles as the type.

713. MOSCHONEURA.

1870. Butl., Cist. Ent. i. 39, 54: **Methymna**, Pinthæus (Pinthœus), Nehemia (Cydno). Methymna specified as type.

714. MURTIA.

1816. Hübn., Verz. 98: **Pyranthe** (Minna). Sole species, and therefore type.

715. MYCALESIS.

1816. Hübn., Verz. 55: **Evadne**, Mineus (Minea, Justina), Mamerta (Hamerta), Medus (Hesione), Ostrea (Otrea).
1844. Doubl., List Br. Mus. 139: employs it for Medus (Hesione), Mineus, Ostrea (Otrea), and some MS. species.
1851. Westw., Gen. Diurn. Lep. 392: the same and others.
1865. Herr.-Schaeff., Prodr. i. 62: the same.
1868. Butl., Cat. Sat. 128: specifies Evadne as type.
1871. Kirb., Syn. Cat. 87: Evadne, Medus, Mineus, Ostrea, and others.

Evadne may be accepted as the type, one at least of the species placed in this group by Doubleday being strictly congeneric therewith. See Orsotriæna.

716. MYLOTHRIS.

1816. Hübn., Verz. 90: **Rhodope** (Arsalte), Ilaire (Margarita), Argia, Lyncida (Monuste, Hippo), Hedyle, Drusilla, Lycimnia (Agrippina), Demophile, Monuste (Hippomonuste).
1850. Steph., Cat. Brit. Lep. 254 [Milothris]: employs it for Monuste only.
1870. Butl., Cist. Ent. i. 34, 42: employs it for Rhodope (Poppea, Rhodope), Agathina, and Trimenia, and specifies Rhodope (Poppea) as type.

It cannot be used for Monuste in accordance with Stephens's usage, since that must be the type of Ascia (q. v.).

717. MYNES.

1832. Boisd., Voy. Astrol. 129: Australis (Leucis), **Geoffroyi**.
1848. Doubl., List Br. Mus. App. 22: employs it for Geoffroyi and others, and therefore this becomes type.
1850. Westw., Gen. Diurn. Lep. 267: follows Boisduval.
1869. Wall., Trans Ent. Soc. Lond. 77: considers Geoffroyi as the type and describes two others.
1871. Kirb., Syn. Cat. 274: follows Wallace.

718. MYRINA.

1807. Fabr., Ill. Mag. vi. 286: **Silenus** (Alcides), Helius (Heleus).
1815. Oken, Lehrb. i. 722: uses it for eight species, specifying only Silenus (Alcides) and Halesus.
1823. God., Encyl. méth. 592: divides the group into two sections, omitting Helius and placing Silenus (Alcides) in the second.
1829. Horsf., Descr. Cat. Lep. E. Ind. Co. 116: employs it for Ravindra and Freja (Jafra), and specifies the latter as the type, erroneously.
1836. Boisd., Spec. gén., pl. 3 B, 6 C.: uses it for the Horsfieldian species and another.
1847. Doubl., List Br. Mus. 21: the same and others.
1852. Westw., Gen. Diurn. Lep. 475: employs it for eighteen species, including Freja and Silenus (Alcides).
1870. Kirb., Journ. Linn. Soc. Lond. x. 500: specifies Silenus (Alcides) as type.

1872. Crotch, Cist. Ent. i. 66: says Silenus (Alcides) is type, through Westwood, 1852.

But it was determined long before that; for Helius was taken in 1816 as type of Iolaus, and nothing but Silenus then remained. See also Loxura.

719. MYSCELIA.*

1844. [Boisd. in] Doubl., List Br. Mus. 88 : Orsis, Numilia (Mycalia), Acontius (Medea).
1849. Boisd. in ib., Gen. Diurn. Lep. 220: employs it for Orsis, Cyaniris, Ethusa, and Antholia.
1861. Feld., Neues Lep. 16: I. Orsis; II. Ethusa, Cyaniris.
1870. Boisd., Lép. Guat. 40: claims it as his own, and says it was founded on the females of Epicalia, and so should be dropped.

The name, too, is unfortunately near Miselia (Ochs., Lep. 1816), and is actually preoccupied through Myscelus (Hübn., Lep. 1816; Heyd., Arachn., 1826). See Sagaritis.

720. MYSCELUS.

1816. Hübn., Verz. 110: **nobilis**, Sebaldus, Erythus.
1852. Westw., Gen. Diurn. Lep. 526: the same and others.
1869. Herr.-Schaeff., Prodr. iii. 58 : uses it for a large number of species, including nobilis and Assaricus.
1869. Butl., Cat. Fabr. Lep. 264: employs it for nobilis and Assaricus.
1870. Butl., Ent. Monthl. Mag. vii. 92: specifies nobilis as the type.
1871. Kirb., Syn. Cat. 587 : uses it in the same sense.

721. NAHIDA.

1871. Kirb., Syn. Cat. App. 651: **cœnoides.** Sole species, and therefore type.

Employed to replace Threnodes preoccupied.

722. NAIS.*

1832–33. Swains., Zoöl. Ill. ii. 136: *Thysbe* (splendens). Sole species, and therefore type.

Subsequently used by Felder. But one of the synonymes of Thysbe is Nais, and the name therefore falls ; it is also preoccupied in Worms (Mull. 1771).

723. NAJAS.

1806. Hübn., Tent. 1: **populi**. Sole species, and therefore type.
    See Nymphalis and Limenitis. See also Nympha, p. 293.

724. NAPÆA.

1825. Hübn., Catal. Franck, 76: Nicæus (Nicæa), Halimede, Eucharila (Actoria), Thersander, Lucinda, Mandana (Mandane), Lucina, Ceneus (Lusca), Athemon (Athemæna), Lamis, Caricæ, Mantus (Mante), Bomilcar (Bombilcar), Phareus (Pharea), Thisbe (Perdita), Pais, Dorilas (Nyx), **Lisias** (Lisiassa), Sagaris.
    Lisias may be taken as the type.

725. NAPEOCLES.

1864. Bates, Journ. Ent. ii. 194: **jucunda**. Sole species, and therefore type.
1872. Kirb., Syn. Cat. 193: the same.

726. NAPEOGENES.

1862. Bates, Linn. Trans. xxiii. 533: I. **Cyrianassa** (Cyrianassa, Tunantina, Adelphe), Inachia (Inachia, Pyrois, Pharo, Ercilla, sulphurina), Ithra, Corena; II. Pheranthes, Crocodes, Duressa. Besides these, not classed in either group, are Tolosa, Larina, Apulia, and Xanthare.
1871. Kirb., Syn. Cat. 24: uses it in the same sense.
    Cyrianassa may be taken as the type.

727. NAROPE.

1849. Boisd. in Doubl., Gen. Diurn. Lep., pl. 50: **Cyllastros**. Sole species, and therefore type.
1851. Boisd. in Westw., Gen. Diurn. Lep. 348: Cyllastros and two others.
    Used in same sense by Herrich-Schaeffer and Kirby.

728. NATHALIS.

1836. Boisd., Spec. gén. 589: **Iole**. Sole species, and therefore type, as stated by Butler.
    Since used in same sense.

729. NECTARIA.

1820. Dalm. in Billb., Enum. Ins. 76: given by Billberg as the equivalent of Fabricius's Idea, which fell, from being founded on the single species **Idea**, which therefore becomes the type of this genus.
    See Idea and Hestia.

730. NECYRIA.

1851. Westw., Gen. Diurn. Lep. 432: **Bellona**, Duellona.
1867. Bates, Journ. Linn. Soc. Lond. ix. 428: the same and others.
1871. Kirb., Syn. Cat. 302: follows Bates.
Bellona, being the species figured, may be taken as the type.

731. NELONE.*

1870. Boisd., Lép. Guat. 23: Mandana (Mandana, Ops), **Fatima** (Fatima, Ovidius), Lucinda, Praxithea, Aurimna.
This name falls before Aphacitis and Emesis.

732. NEMEOBIUS.

1827. Steph., Ill. Brit. Ent. Haust. i. 28: **Lucina**. Sole species, and therefore type.
1847. Doubl., List Br. Mus. 2: the same and another.
1851. Westw., Gen. Diurn. Lep. 419: Lucina.
1871. Kirb., Syn. Cat. 284: the same.

733. NEOMÆNAS.

1858. Wallengr., K. Vet. Akad. Förh. xv. 78: **Servilea**. Sole species, and therefore type.
But the species was inedited at this time, being first published, under the same generic name, two years later.

734. NEONYMPHA.

1816. Hübn., Verz. 65: **Phocion** (Helicta), Cornelius (Gemma), Clorimena (Clerimene), Dorothea.
1844. Doubl., List Br. Mus. 137: employs it for Phocion (areolatus), Cornelius (Gemma), and two others.
1851. Westw., Gen. Diurn. Lep. 374: uses it for twenty-five species, among them the two of Hübner's mentioned by Doubleday.
1865. Herr.-Schaeff., Prodr. i. 59: makes a similar use of it.
1868. Butl., Cat. Sat. 35: specifies Phocion (Helicta) as type.

735. NEOPE.*

1867. Butl., Ann. Mag. Nat. Hist. [3] xix. 166: **Bhadra**, Pulaha, Goschkevitschii (Moorei, japonica).
1868. Ib., Ent. Monthl. Mag. iv. 195; Cat. Sat. 112: specifies Bhadra as type.
The name was proposed to supplant Enope preoccupied; but it is itself preoccupied through Neopus (Hodgs., Birds, 1844). Probably some English entomologist (the passion for the formation of generic terms by transposition seems to be strong in England) will propose Nopea, Epone, Opene, or Onepe, to take its place!

736. NEOPHASIA.

1869. Behr, Trans. Am. Ent. Soc. ii. 303 : **Menapia**, Terlooii.
Menapia may be taken as the type.

737. NEORINA.

1850. Westw., Gen. Diurn. Lep., pl. 65 : **Hilda**. Sole species, and therefore type, as stated by Butler.
1851. Ib., ib. 369 : the same.

738. NEORINOPIS (fossil).

1873. Butl., Lep. Exot. i. 127 : **sepulta**. Sole species and designated type.

739. NEOSATYRUS.

1858. Wallengr., K. Vet. Akad. Förh. xv. 79 : **ambiorix**. Sole species, and therefore type, as stated by Butler.

740. NEPHERONIA.

1870. Butl., Cist. Ent. i. 38, 53 : **Poppea** (Idotea), Thalessina, Argia, Buquetii, Pharis (Pharis, Chione), Valeria (Hippia, Bœbera), Iobœa (Jobœa). Poppea designated as type.

741. NEPTIS.

1807. Fabr., Ill. Mag. vi. 282 : **Melicerta**, aceris.
1816. Hübn., Verz. 42 : Nauplia, Emilia. These have nothing to do with the Fabrician group.
1820. Billb., Enum. Ins. 78 : changes the name, for no reason, to Philonoma (q. v.).
1828. Horsf., Descr. Cat. Lep. E. Ind. Co., expl. pl. 5, 7 : Vikasi? populi, and aceris.*
1850. Westw., Gen. Diurn. Lep. 270 : divides the group into two sections, and places both Fabricius's species in the second.
1861. Feld., Neues Lep. 30 : divides the group into seven sections, placing aceris in the first, with others, and Melicerta alone in the fourth.
1872. Crotch, Cist. Ent. i. 66 : says aceris is type, through Horsfield and Westwood ; but Westwood's action certainly has no effect upon it.

Aceris, however, has belonged to Nymphalis (q. v.) since 1823 ; and hence Melicerta must be taken as the type of this group.

---

* He has also referred aceris, in the same work, to a new (unnamed) genus.

742. NEREIS.*
1806. Hübn., Tent. 1: *Polymnia.* Sole species, and therefore type.
The name is preoccupied in Worms (Linn. 1731). See Mechanitis.

743. NERIAS.
1836. Boisd., Spec. gén., pl. 4 A., 7 B.: Calliope, Euterpe, **Susanna.**
The larva and pupa only of the first two are figured.
1844. Doubl., List Br. Mus. 64: employs it for Phlegia and Susanna only.
Susanna, therefore, becomes the type. The name is very near to Neria (Rob.-Desv., Dipt. 1830).

744. NESSÆA.*
1816. Hübn., Verz. 41: Obrinus (Ancæa), Harpalyce, Galanthis.
This name is preoccupied through Nesæa (Lamx., Pol. 1812).

745. NESTORIDES.
1816. Hübn., Verz. 86: **Gambrisius** (Drusius, Gambrisius, Amphitrion). Sole species, and therefore type.

746. NETROCORYNE.
1867. Feld., Reise Novara, 507: **repanda.** Sole species, and therefore type, as stated by Butler.
1867. Hewits., Hundr. Hesp. 22: beata, Denitza.
1869. Herr.-Schaeff., Prodr. iii. 69: repanda, cœcutiens.
1871. Kirb., Syn. Cat. 621: all the above.

747. NEUROSIGMA.
1868. Butl., Proc. Zoöl. Soc. Lond. 615. **Siva** specified as type.
1871. Kirb., Syn. Cat. 249: the same.
See Acontia.

748. NICA.*
1822-26. Hübn., Exot. Schmett. ii.: *flavilla.* Sole species, and therefore type.
Since used in the same sense by Doubleday, Felder, and Kirby; but the name is preoccupied through Nika (Risso, Crust. 1816).

749. NICONIADES.
1816-21. Hübn., Exot. Schmett. ii.: **Xanthaphes** (Xanthaptes). Sole species, and therefore type.
1821. Ib., Index, 7: the same.
This name is too close to Nisoniades of the same family, proposed by the same author (Verz. 1816), to stand. It cannot, however, have been a simple typographical error. See Goniloba.

750. NINA.*

1829. Horsf., Descr. Cat. Lep. E. Ind. Co. 140: *Xiphia* (Nina). Sole species, and therefore type.

The name is one of the synonymes of Xiphia, and therefore falls. See Leptosia and Nychitona.

751. NIRODIA.

1851. Westw., Gen. Diurn. Lep. 430: **Belphegor.** Sole species, and therefore type.

752. NISONIADES.

1816. Hübn., Verz. 108: **Bromius,** Mimas, Zephodes, Juvenalis (Juvenis), Tages, Flesus (Ophion), and a MS. species.
1850. Steph., Cat. Brit. Lep. 22: restricts the name to Tages, but this had already (1832) been taken to form Thanaos.
1852. Westw., Gen. Diurn. Lep. 579: employs it for all of Hübner's species excepting Zephodes and Flesus, and for many others.
1869. Butl., Cat. Fabr. Lep. 286: employs it for Tages, Juvenalis, Mimas, and others.
1871. Staud., Cat. Eur. Lep. 34: uses it for Tages and others.

Other authors have used it similarly. Bromius may be taken as the type. See Thanaos.

753. NOMIADES.

1816. Hübn., Verz. 67: **Semiargus** (Acis), Atys (Pheretes), Damon, Cyllarus (Damœtas), Arcas (Erebus), Alsus, Alcon, Diomedes (Euphemus), Arion, Lysimon.
1850. Steph., Cat. Brit. Lep. 19, 261: employs the name for Alsus, Semiargus (Acis), Arion, and Alcon.
1858. Kirb., List Brit. Rhop.: uses it for the same, excepting Alsus, and for others. The query attached to many of them only indicates that they are doubtful as British species.

Semiargus may be taken as the type.

754. NOTHEME.

1851. Westw., Gen. Diurn. Lep. 462: **Eumeus** (Ouranus). Sole species, and therefore type.
1867. Bates, Journ. Linn. Soc. Lond. ix. 425: the same and another.

See Amblygonia.

755. NYCHITONA.*

1870. Butl., Cist. Ent. i. 34, 41: *Alcesta* (Dorothea), Xiphia (Niobe). The former specified as type.

The name falls before Leptosia. See also Nina.

756. NYMPHALIS.

1805. Latr., Sonn. Buff. xiv. 82: (*nymphales*) Jason (Jasius), Antiopa, Polychloros, vau. album (v. album), urticæ, c. album, Egea (triangulum), Levana (Prorsa, Levana), Atalanta, Io, cardui, celtis, populi, **Sappho** (Lucilla), Sibylla (Sibilla), Iris (Iris, Beroe); (*perlati*) Paphia (Paphia, Valesiana), Adippe, Aglaia, Daphne, Dia, Pales, Euphrosyne (Euphrosine), Lathonia, Lucina, Cynthia, Aurinia (Artemis), Cinxia; (*satyri*) Circe, Hermione, Briseis, Fidia, Statilimus (Fauna), Actæa, Semele, Phædra, Ligea, Melampus, Manto (Pollux), Medea (Æthiops), Dejanira, Ægeria, Mæra (Satyrus), Hyperanthus, Tithonus (pilosellæ), Jurtina (Janira), Pamphilus, Arcania (Arcanius), Galathea.

As Latreille indicates the first of these groups as typical by giving it the distinctive name *nymphales*, any further restriction of the genus must be confined to this group.

1810. Ib., Consid. 440: Dido, aceris, populi, and Achilles are specified as types.

Populi is the only one given in the previous list, none of the others being even congeneric with any of the species then referred to the genus. This, therefore, would become the type, had it not been previously taken as the type of Najas (q. v.). Latreille's treatment of the group in his Genera (1809), and in Cuvier's Règne Animal (1817), is essentially the same as in Sonnini's Buffon.

1816. Lam., Hist. Nat. An. sans Vert. iv. 24: employs it for the Satyrids only, but of course erroneously.

1823. God., Tab. Meth. 43: uses it for Jason (Jasius), Iris, Ilia, populi, Sibylla, Camilla, Sappho (Lucilla).

Jason had already been taken as the type of Charaxes, as Felder has pointed out; Iris, and consequently Ilia, were removed to Potamis in 1806. Camilla became the type of Limenitis in 1815, taking with it Sibylla; so that Sappho must be considered the type of this genus.

1828. Boit., Man. Ent. ii. [Nymphalus].
1829. Boisd., Index, 16: restricts it to populi.
1832. Dup., Pap. France, Diurn. Suppl. 401: the same.
1833. Brullé, Exp. Morée, 283: uses it for Jason (Jasius) only.

1844. Doubl., List Br. Mus. 96: employs it for populi, Artemis, etc.
1850. Westw., Gen. Diurn. Lep. 306: considers Jason as type.
1861. Feld., Neues Lep. 41: divides the group into four sections, including the genera Cymatogramma and Paphia of Doubleday's Genera, but mentions no species referred to the group by Latreille. See remarks in his note.
1871. Kirb., Syn. Cat. 267; uses it for Jason and allies; but on p. 648 makes it supplant Vanessa, and refers the genus to Linné.
1872. Crotch, Cist. Ent. i. 60: also wrongly refers the genus to Linné [Nymphales], and says that Lamarck in 1801 (where only the plural form is used) fixed the type as Atalanta.
1872. Scudd., Syst. Rev. 10: specifies Polychloros as type, erroneously. See also Neptis and Limenitis.

757. NYMPHIDIUM.

1807. Fabr., Ill. Mag. vi. 286: Caricæ, **Thelephus** (Telephus), Athemon.
1815. Oken, Lehrb. i. 722: the same and others.
1832. Boisd.-LeC., Lép. Am. Sept. 130 [Nymphidia]: Cæneus (Pumila).
1836. Boisd., Spec. gén., pl. 2 B.: Jessa.
1847. Doubl., List Br. Mus. 10: employs it for a large number, including, of Fabricius's species, only Caricæ.
1851. Westw., Gen. Diurn. Lep. 447: employs it for twenty species, including Caricæ and Lamis, which are specified as "representative."
1868. Bates, Journ. Linn. Soc. Lond. ix. 450: uses it for nearly fifty species, including Caricæ.
1872. Crotch, Cist. Ent. i. 66: says that Caricæ is type, through Westwood in 1850 [1851].

Caricæ, however, belongs to Peplia, 1816, and Athemon to another family. Hence Thelephus should be taken as the type. See Peplia, Desmozona, and Heliochlæna.

758. NYMULA.

1836. Boisd., Spec. gén., pl. 4 C.: **Gnosis**. Sole species, and therefore type.
1840. Blanch., Hist. Nat. Ins. iii. 465 [Nimula].
1868. Bates, Journ. Linn. Soc. Lond. ix. 450: employs it for Gnosis and allies.

759. Nytha.

1820. Billb., Enum. Ins. 77: Hyperbius, Medea, Clytus, Mæra, Agave (Alcyone, Hippolyte), Semele, **Hermione**, Briseis, Statilimus (Faunus), Fidia, and several MS. Species.

All these species, excepting Hermione and Briseis, fall into the genera Erebia, Eumenis, Pararge, Melampias, and Dira; all of which are of earlier date. The name may therefore be retained for these two species and their allies, with Hermione for type. See also Maniola and Oreas.

760. Oarisma.

1872. Scudd., Syst. Rev. 54: **Poweshiek.** Sole species and designated type.

761. Ocalis.*

1851. Boisd. in Westw., Gen. Diurn. Lep. 371: Westwood gives it as a MS. synonyme of Oressinoma (q. v.).

1870. Boisd., Lép. Guat. 63: *Typhla.* Sole species, and therefore type.

This name falls before Oressinoma; and is besides too close to Ocalea (Erichs., Col. 1837).

762. Ochlodes.

1872. Scudd., Syst. Rev. 57: **nemorum**, agricola, Sonora. Nemorum specified as type.

763. Ocytes.

1872. Scudd., Syst. Rev. 55: **Metea**, Seminole. Metea specified as type.

764. Œneis.

1816. Hübn., Verz. 58: **Norna** (Norna, Celæno), Polixenes (Bore), Jutta, Arethusa.

1868. Butl., Ent. Monthl. Mag. iv. 196; Cat. Sat. 160: specifies Norna as the type.

1871. Staud., Cat. 27: uses it in the same sense.

1871. Kirb., Syn. Cat. 68: the same.

1872. Scudd., Syst. Rev. 4: specifies Norna as type.

See also Chionobas.

765. Œnomaus.

1816. Hübn., Verz. 76: Marsyas, **Ortygnus**, Eumolphus, Rustan, Palegon.

1869. Butl., Cat. Fabr. Lep. 196: employs it for Marsyas.

But this is already the type of Pseudolycæna. Ortygnus may be chosen as the type.

766. OGYRIS.

1847. Doubl., List Br. Mus. 20: Idmo, **Abrota**, Damo (all inedited).
1852. Westw., Gen. Diurn. Lep. 472: Abrota, Idmo: the former is figured.
1871. Kirb., Syn. Cat. 425: Abrota, Idmo, and others.

Since Doubleday's genus was undescribed, and at the time when it was proposed all the species were inedited, the genus can only date from 1852, though it should bear Doubleday's name: at this time the only published species was Abrota, and this therefore must be the type. Idmo was not published until 1862, and Damo is still a MS. name.

767. OILEIDES.

1822–26. Hübn., Exot. Schmett. ii.: Vulpinus, **Zephodes**.
Zephodes may be taken as the type.

768. OLERIA.

1816. Hübn., Verz. 9: **Astrea**, Flora, Aegle (Clio).
1862. Bates, Linn. Trans. xxiii. 529: Phyllodoce, Theaphia: these species are allied to Hübner's, but have no immediate connection with them. See Scada.
1864. Herr.-Schaeff., Prodr. i. 47: follows Bates.
Astræa may be taken as the type.

769. OLIGORIA.

1872. Scudd., Syst. Rev. 61: **maculata**. Sole species and designated type.
Is this name too close to Oligorus (Dej., Col. 1833)?

770. OLINA.*

1848. Doubl., Gen. Diurn. Lep., pl. 31: *Azeca*. Sole species, and therefore type.
1851. Westw., Gen. Diurn. Lep. 407: Azeca, Emilia.
The name is preoccupied in Diptera (Rob.-Desv. 1830). See Vila.

771. OLYNTHUS.

1816. Hübn., Verz. 80: Inachus, **Narbal**.
Narbal may be taken as the type.

772. OLYRAS.*

1847. Doubl., Gen. Diurn. Lep. 107: *Crathis*. Sole species, and therefore type.
Used in same sense by Herrich-Schaeffer and Kirby, but the name is preoccupied through Olyra (McClell., Fishes, 1842), and perhaps through Oluris (Heyd., Arachn. 1826).

773. OPSIPHANES.

1849. Doubl., Gen. Diurn. Lep., pl. 57: Boisduvalii, **Sallei**, Reevesii.
1851. Westw., Gen. Diurn. Lep. 344: Syme and eleven others, including Boisduvalii, Sallei, Xanthus, and Berecynthus, but not Reevesii. Xanthus and Berecynthus are specified as types, but they cannot be, because they were not of the original species.
1864. Herr.-Schaeff., Prodr. i. 54: Berecynthus and others.
1871. Kirb., Syn. Cat. 125: all the above but Reevesii.

Sallei may be taken as the type.

774. OREAS.*

1806. Hübn., Tent. 1: *Circe* (Proserpina). Sole species, and therefore type.
1815. Oken, Lehrb. i. 740: the same and others.
1865. Feld., Reise Novara, 305: Marathon, Ctesiphon: wholly unrelated to the above. See Rusalkia.
1867. Bates, Journ. Linn. Soc. Lond. ix. 431: follows Felder.

The name is preoccupied in Mammals (Desm. 1804). See Nytha.

775. OREINA.*

1840. Westw., Brit. Butt. 76: Ligea, Medea (Blandina), *Epiphron* (Cassiope).
1867. Butl., Ent. iii. 277: Epiphron and others not in preceding list.
1868. Ib., Ent. Monthl. Mag. iv. 194: specifies Epiphron (Cassiope) as type.

The name is preoccupied in Coleoptera (Chevr. 1834). See Erebia, Gorgo, Marica, Syngea, Phorcis, and Epigea.

776. ORESSINOMA.

1850. Westw., Gen. Diurn. Lep., pl. 62: **Typhla**. Sole species, and therefore type, as stated by Butler.
1851. Ib., ib. 371: the same.

See Ocalis.

777. ORESTIAS.*

1862. Feld., Wien. Ent. Monatschr. vi. 73: *Vitula*. Sole species, and therefore type.
1867. Bates, Journ. Linn. Soc. Lond. ix. 432 [Orestia]: the same and another.

1868. Herr.-Schaeff., Prodr. iii. 7 [Orestia]: follows Bates.

The name is preoccupied in Fishes (Val. 1839) and also through Orestia in Coleoptera (Chevr. 1834). See Cartea.

778. ORIMBA.

1856. Boisd. in Herr.-Schaeff., Exot. Schmett. 55: **Cepha** (Cataleuca), Pasiphae (Arcas).

Pasiphae was taken in 1847 as the type of Pandemos, so that Cepha must be taken as the type of this genus. See Aricoris, in founding which Boisduval also made use of Cepha!

779. ORINOMA.

1846. Doubl., in Gray's Nepaul, 14: **Damaris**. Sole species, and therefore type, as stated by Butler.

Used for this species only by Westwood, Herrich-Schaeffer, Butler, and Kirby.

780. ORNITHOPTERA.

1832. Boisd., Voy. Astrol. 33: **Priamus**, Helena (Amphimedon, Helena).
1836. Ib., Spec. gén. 173: Priamus, Helena, and others.

Used in same sense by Doubleday and Westwood. Priamus may be taken as the type.

781. ORPHEIDES.

1816. Hübn., Verz. 86: **Demoleus**, Erithonius (Epius).

Demoleus may be taken as the type.

782. ORSOTRIÆNA.*

1858. Wallengr., K. Vet. Akad. Förh. xv. 79: *Medus* (Hesione). Sole species, and therefore type.

The name probably falls before Mycalesis.

783. OXEOSCHISTUS.

1867. Butl., Ann. Mag. Nat. Hist. [3] xx. 268: **Puerta**, Hilara, Protogenia, Pronax, Propylea, Prochyta, Irmina, Tauropolis. Puerta specified as type.
1868. Ib., Cat. Sat. 180: the same.
1871. Kirb., Syn. Cat. 106: follows Butler.

784. OXYLIDES.

1816. Hübn., Verz. 77: Celmus, **Faunus**.

Faunus may be taken as the type.

785. OXYNETRA.

1862. Feld., Wien. Ent. Monatschr. vi. 179: **semihyalina**. Sole species, and therefore type.
1871. Kirb., Syn. Cat. 583: the same.

786. PACHLIOPTERA.

1864. Reak., Proc. Ent. Soc. Phil. iii. 503: Darsius, Pompeus, Rhadamanthus, Priamus, Hector, **Aristolochiæ** (Diphilus), Philenor? Polydamus? Clytia (dissimilis).
 Aristolochiæ may be taken as the type. See also Polydorus.

787. PACHYRHOPALA.*

1858. Wallengr., K. Vet. Akad. Förh. xv. 81: *Phidias*. Sole species, and therefore type.
 The name falls before Tamyris.

788. PACHYTHONE.

1867. Bates, Journ. Linn. Soc. Lond. ix. 389: **Erebia**, Lateritia, distigma, Xanthe, mimula.
1871. Kirb., Syn. Cat. 316: the same.
 Erebia may be taken as the type.

789. PAGYRIS.*

1870. Boisd., Lép. Guat. 34: *Ulla*. Sole species, and therefore type.
 Preoccupied through Pagurus (Fabr., Crust. 1798).

790. PALÆONTINA* (fossil).

1873. Butl., Lep. Exot. i. 126: *oolitica*. Sole species, and therefore type.
 It is not a butterfly, as I shall show in my memoir on Fossil Butterflies, shortly to be published by the Amer. Assoc. Adv. Science.

791. PALLA.

1816. Hübn., Verz. 47: **Decius** (Decia). Sole species, and therefore type. Used by Kirby (Syn. Cat.) in same sense.
 See also Phyllophasis and Philognoma.

792. PALLENE.*

1848. Doubl., Gen. Diurn. Lep., pl. 41: *Eupithes*. Sole species, and therefore type.
1850. Westw., Gen. Diurn. Lep. 289: the same.
 The name is preoccupied in Coleoptera (Meg. 1823; Lap. 1836), in Crustacea (Johnst. 1837), and in Birds (Less. 1837).

793. PAMPHILA.

1807. Fabr., Ill. Mag. vi. 287: comma, **Palæmon** (Paniscus), malvæ (Fritillum, lavateræ).
1828. Steph., Ill. Brit. Ent. Haust. i. 99: employs it for Palæmon (Paniscus), comma, and others, placing malvæ elsewhere.
1837. Curtis, Guide, 2d Ed. 174: makes similar use of it.
1840. Westw., Gen. Syn. 88: specifies comma as type.
1840 [ined.?] Ramb., Faune Ent. Andal. 321: malvæ, Proto, etc.
1858. Ib., Cat. Lép. Andal. 78: uses it for Proto and others.
1858. Kirb., List Brit. Rhop.: employs it for sylvanus only, a species not mentioned by Fabricius.
1869. Butl., Cat. Fabr. Lep. 276: uses it for several species, including only comma of Fabricius.
1870. Butl., Ent. Monthl. Mag. vii. 93: specifies comma as type.
1871. Kirb., Syn. Cat. 596: employs it in a very wide sense, including comma.
1872. Crotch, Cist. Ent. i. 67: says that comma is type, through Dalman's action in 1816. But Dalman did not use the name Pamphila even as a synonyme!
1872. Scudd., Syst. Rev. 56: specfies comma as type.

Comma, however, cannot be taken as the type, for in 1832 it virtually became the type of Erynnis (q. v.); malvae already belonged to Hesperia in 1798; and therefore Palaemon must be taken as the type. See Carterocephalus and Steropes.

794. PANARA.

1847. Doubl., List Br. Mus. 8: Sagaris (Satnius), **Thisbe** (Iarbas).
1851. Westw., Gen. Diurn. Lep. 442: employs it for Thisbe (Iarbas), Phereclus (Barsacus), and Sagaris (Satnius), the last with a query.

Thisbe should therefore be considered the type; and in this sense it is used by Bates and Kirby.

795. PANDEMOS.

1816. Hübn., Verz. 25: Placidia, Liberia, **Pasiphae** (Arcassa), Lagus (Lagis).
1847. Doubl., List Br. Mus. 7: employs it for Pasiphae (Arcas) and others.
1851. Westw., Gen. Diurn. Lep. 440: does the same, and specifies Pasiphae (Arcas) as type.

1867. Bates, Journ. Linn. Soc. Lond. ix. 456: uses it for Pasiphae (Arcas) only.
1871. Kirb., Syn. Cat. 332: follows Bates.

796. PANDITA.

1858. Moore, Cat. Lep. E. Ind. Co. i. 181: **Sinope**. Sole species, and therefore type.

Used in same way by Felder and Kirby.

797. PANDORA.*

1848. Boisd. in Doubl., Gen. Diurn. Lep., pl. 43: *Prola*. Sole species, and therefore type.
1850. Boisd. in Westw., Gen. Diurn. Lep. 300: the same.

The name is subsequently used by Felder; but it is preoccupied in Mollusks (Brug. 1791; Meg. 1811), in Acalephs (Eschsch. 1829), and in Diptera (Halid. 1833). See Batesia.

798. PANOPEA.*

1816. Hübn., Verz. 39: Semire, Lucretia.
1850. Westw., Gen. Diurn. Lep. 281: employs it for the same and another.
1861. Feld., Neues Lep. 27 [Panopæa]: description only.
1865. Herr.-Schaeff., Prodr. i. 65 [Panopæa]: uses it for Hübner's species and others.
1869. Butl., Cat. Fabr. Lep. 95: the same.

But the name is preoccupied through Panopæa (Mén., Moll. 1807). See Pseudacræa.

799. PANOPLUIA.*

1864. Reak., Proc. Ent. Soc. Phil. ii. 503: no species mentioned.

Indeed, it is established as an hypothetical genus, for a form of Papilionides, with an anopluriform larva, yet to be discovered!! Credat Judæus Apella!

800. PANSYDIA.

1872. Scudd., Syst. Rev. 60: **Cunaxa** (Cunaxa, Mesogramma). Type specified as Cunaxa (Mesogramma).

801. PANTHIADES.

1816. Hübn., Verz. 79: **Pelion** (Thallus, Pelion). Sole species, and therefore type.
1869. Butl., Cat. Fabr. Lep. 197: employs it for Pelion and five others.

802. PANTOPORIA.

1816. Hübn., Verz. 44: Phærusa, Nefte, **Hordonia**, Dorcas (Mardania).

Hordonia may be taken as the type.

803. PAPHIA.*

1807. Fabr., Ill. Mag. vi. 282: I. *Jason* (Jasius), Pollux; II. Varanes, Morvus (Laertes), Chorinæus; III. Medon, Astyanax (Ursula); IV. Odius (Orion), Isidora (Itys), Acontius (Antiochus).

1829. Meig., Eur. Schmett. i. 95: uses it for Jason (Jasius) only, which therefore becomes type, as stated by Crotch (Cist. Ent. i. 66).

It has subsequently been used in same sense by many naturalists, but is preoccupied in Mollusks (Lam. 1801). See Charaxes and Jasia.

804. PAPILIO.

[1735. Linn., Syst. Nat. ed. i. (Fee's Reprint, p. 76): no species mentioned; intended to include all Lepidoptera, divided into three groups, of which butterflies form the first.

1736. Ib., Acta Upsal. iv. 117: species mentioned (without names) are, as given by Hagen:* rhamni, brassicæ, rapæ, napi, cratægi, Apollo, Antiopa, polychloros, urticæ, c. album, Io.

1740. Ib., Syst. Nat. ed. ii. 60: no species mentioned; divided into several groups by the structure of the antennæ and mouth parts, those "pedibus 4," *i.e.* Nymphales, placed first. Essentially the same arrangement occurs in the third, fourth, and fifth editions.

1746. Ib., Faun. Suec. ed. i. 232: the butterflies are divided into two groups, according as they have four (serviceable) or six legs. Species occur again without names, but numbered from 772 to 807 inclusive; 772 was afterwards named Antiopa.

1748. Ib., Syst. Nat. ed. vi. 63: species are introduced for the first time in a general work, but still unnamed. Sixteen butterflies only are mentioned, all of them before treated of in the Fauna Suecica. Antiopa heads the list. The arrangement of the seventh and ninth edition is identical; the eighth edition contains no animals.

I have introduced the foregoing only for its historic interest. The reader will find fuller details in papers by Dr. Hagen and myself in the Canadian Entomologist, vol. vi. pp. 143–145, 163–166. In this matter I hold to the views of Agassiz, who lays down as a maxim for genera: *Cum binominalis nomenclaturæ Linnæus sit auctor, illa de prioritatu lex ad anteriorum auctorum opera non est retorquenda.*† I do not therefore deem even Linné's action (previous to 1758, when binomial nomenclature was founded) to have had any binding force; yet, in view of the opinions I expressed in my Systematic Revision (p. 16), without examination of

---

* Can. Ent. vi. 165.   † Nomencl. Zoöl. Introd. xx.

Linné's action previous to 1758, it is interesting to discover that, in the first mention of species under Papilio, Antiopa is introduced, and Machaon (or any other swallow-tail) is not; further, that Antiopa is mentioned in every subsequent use of the generic name by Linné, and that, in every instance, excepting in 1736, when species are first referred to, the group to which Antiopa belongs, or, if species are directly mentioned, this species itself, stands first upon the list, as if that insect, at least, were always in his thought when Papilio was recorded. I repeat, however, that this consideration has no binding force whatsoever.]

1758. Ib., Syst. Nat. ed. x. i. 458: employs it for all butterflies then known, which are now described in full, and supplied with binomial nomenclature for the first time. They are divided as follows: Equites (Trojani, Achivi), Heliconii, Danai candidi, Danai festivi, Nymphales (gemmati, phalerati), Plebeii (rurales, urbicolæ), Barbari. Among the butterflies occur **Antiopa**, Machaon, Podalirius, and Memnon.*

1793. Fabr., Ent. Syst. iii. 1, 258: removes from this great group all the Rurales and Urbicolæ, under the name of Hesperia, thus confining Papilio to the Nymphales and Papilionides.

1801. Schrank, Faun. Boica, ii. i. 152, 188: restricts the name still further to the Nymphales, and divides the group, thus limited, into sections, as follows: † I. populi (Semiramis, populi), Sibylla (Sibilla), Camilla, aceris (Lucilla); II. Atalanta, cardui, Io (Jo), Antiopa, Polychloros, urticæ, c. album, Levana (Prorsa, Levana); III. Paphia, Adippe (Syrinx, Adippe), Niobe, Aglaia (Agluja), Lathonia, Dia, Selene (Thalia); IV. Athalia (Phœbe), Maturna, Cynthia (Cinthia), Hecate, Ino (Dictynna), Aurinia (Artemis), Didyma (Cinxia), Cinxia (Trivia), Lucina.

1805. Latr., Sonn. Buff. xiv. 108: first restricts the name to the "swallow-tails," but erroneously, as these had already been excluded by Schrank's limitation.

In this action he has been followed by all authors for nearly seventy years, until now it has become the all but universal custom to apply it to an immense group of over three hundred species, really composed of a vast number of genera, as any one may judge by a comparison of their earlier stages, which show greater differences than can be found in any other generally accepted genus of butterflies. Compare also Felder's study of this great group.

---

* I specify these, on account of what follows.
† All the names are used by Linné.

1810. Latr., Consid. 440: specifies Machaon as the type, but, of course, erroneously.
1815. Oken, Lehrb. i. 722: makes this one of his groups of Emesis, and refers to it Melander (one of the Vestales). See also Pieris.
1816. Hübn., Verz.: makes no use of it whatever.
1832-33. Swains. Zoöl. Ill. ii. 95: considers Memnon a "pre-eminently typical" species.
1836. Curtis, Brit. Ent. pl. 578: specifies Podalirius as the type.
1840. Westw., Gen. Syn. 87: specifies Machaon as the type.
1864. Reak., Proc. Ent. Soc. Phil. ii. 60, 62: separates the swallow-tails into several genera, retaining Papilio for one of the groups, including Memnon, Machaon, Thoas, Glaucus (Turnus), etc.
1872. Crotch, Cist. Ent. i. 60: says that Cuvier, in 1799, marked Machaon as the type.

But nothing can be found in the Tableau Élémentaire to warrant such a statement. Cuvier places all the butterflies under Papilio, dividing it into sections, to which the names Nymphales, Danai, Parnassii, Heliconii, Equites, and Plebeii are given; and under Equites he gives "P. Machaon" as an example or type. Certainly, from the contents of Cuvier's work, there is no more reason for selecting this as type than "P. Antiopa," which is the first example given under the Nymphales.

1872. Scudd., Syst. Rev. 11: specifies Antiopa as the type.

See Scudderia, Amaryssus, Princeps, Iphiclides, and Iliades.

805. PARAMACERA.

1868 (Feb.). Butl., Ent. Monthl. Mag. iv. 194: Conhiera (a MS. species) given as type. No species whatever are described.
1868. Ib., Cat. Sat. 98: **Xicaque.** Sole species, and therefore type.

Probably Butler found his Conhiera to be synonymous with Reakirt's Xicaque, and therefore simply suppressed his own name; but it would have been well to have simply stated the fact, if it were so.

806. PARAMIMUS.

1816. Hübn., Verz. 115: **Scurra,** Talaps, Eumelus.
1869. Herr. Schaeff., Prodr. iii. 52: without mention of species.
1870. Butl., Ent. Monthl. Mag. vii. 97: specifies Scurra as type.

807. PARAPLESIA.*

1862. Feld., Wien. Ent. Monatschr. vi. 26: *Adelma.* Sole species, and therefore type.

According to Felder, this name is preoccupied. See Isodema.

808. PARARGE.

1816. Hübn., Verz. 59: **Aegeria** (Egeria, Xyphia). Sole species, and therefore type, as stated by Butler, and as used by Stephens, Rambur, Heydenreich [Pararga], Staudinger [Pararga, 1861; Pararge, 1871], and Herrich-Schaeffer [Pararga].

809. PARDALEODES.

1870. Butl., Ent. Monthl. Mag. vii. 96: **Edipus**, Laronia. Edipus specified as type.
1871. Kirb., Syn. Cat. 625: uses it in the same way.

810. PAREBA.

1848. Doubl., Gen. Diurn. Lep. 142: **vesta**. Sole species, and therefore type.

811. PARIDES.

1816. Hübn., Verz. 87: **Echelus**, Æneas, Æneides (Gargasus), Anchises (Lysander), Vertumnus, Sessostris.
Echelus may be taken as the type.

812. PARNASSIUS.

1805. Latr., Sonn. Buff. xiv. 110: **Apollo**, Mnemosyne, Polyxena (Hypsipyle), Rumina.
1810. Ib., Consid. 440: Apollo specified as type.
1815. Oken, Lehrb. i. 725: uses it for Apollo and others.
1816. Hübn., Verz. 90: [Parnassis], Apollo, Phœbus (Delius), Mnemosyne.
1816. Lam., Hist. Nat. An. sans Vert. iv. 32: Apollo, Mnemosyne.
1844. Doubl., List Br. Mus. 21: Apollo and others.
1847. Ib., Gen. Diurn. Lep. 26: the same.
1864. Feld., Spec. Lep. 39: divides the species, twenty-seven in number, into two sections, placing Mnemosyne in the first and Apollo in the second.
See Doritis and Therius.

813. PARNES.*

1847. Doubl., List Br. Mus. 18: Nycteis, Philotes (both unpublished species).
1851. Westw., Gen. Diurn. Lep. 464: the same. Both are described, and Nycteis figured.
1867. Bates, Journ. Linn. Soc. Lond. ix. 436: the same.

1871. Kirb., Syn. Cat. 111: the same.
>If the name could be retained, it should be with Doubleday's name, and the date 1851; and Nycteis could most appropriately be taken as type; but it is preoccupied through Parnus (Fabr., Col. 1792).

814. PAROMIA.*
1861. Hewits., Exot. Butt. ii.: *pulchra*. Sole species, and therefore type.
>The name is preoccupied in Coleoptera (Westw. 1850).

815. PARRHASIUS.
1816. Hübn., Verz. 79: Timoleon, Arogenus, Hemon (Acmon, Hemon), **Polibetes** (Polybetes), Lisus.
>Polibetes may be taken as the type.

816. PARTHENOS.
1816. Hübn., Verz. 38: **Sylvia.** Sole species, and therefore type.
1871. Kirb., Syn. Cat. 230: the same and others.
>See Minetra.

817. PATHYSA.
1864. Reak., Proc. Ent. Soc. Phil. iii. 503: Sarpedon, Agamemnon, Eurypylus, **Antiphates,** Ajax (Marcellus, Ajax), Podalirius? Sinon.
>Antiphates may be taken as the type.

818. PAVERMIA.*
1864. Reak., Proc. Ent. Soc. Phil. iii. 503.
>This is another of Reakirt's astonishing hypothetical genera, established upon supposititious forms of swallow-tails, with "vermiform" larvæ, yet to be discovered!

819. PAVONIA.*
1823. God., Encycl. méth. Suppl. 807: Actorion, Aorsa, Automedon, Eurylochus, Ilioneus, Teucer, Idomeneus, Demosthenes (Inachis), Martia, Taramela, Batea (Saronia), Syme (Acadina), Rusina, Creusa (Anaxandra), Darius (Anaxerete), Hercyna (Anosia), Darius, Œthon, Berecynthus, Xanthus, Cassiope (Caryatis), cassiæ.
>The name has since been used by several authors, but is preoccupied in Polyps (Lam. 1816), as well as in Lepidoptera (Hübn. 1816).

820. PEDALIODES.
1867. Butl., Ann. Mag. Nat. Hist. [3] xx. 267: **Poesia,** Proerna, Pisonia (Pisonia, dejecta), Perperna, Phanias, Paneis,

Polusca, Pausia, Piletha, Prytanis, Phœnissa, Physcoa, Porina, Peucestas, Pallantis, Pylas, Plotina, Parepa Phila, Phæa, Peruda, Panyasis, Napæa. Poesia specified as type.
Subsequently used by Butler and Kirby.

821. PELEUS.*

1832–33. Swains., Zoöl. Ill. ii. 75: Dan (Eacus), Gentius, *Peleus*. Peleus specified as type.
The name is founded on that of one of the species upon which the genus is founded, and therefore falls. It is also preoccupied through Pelias (Merr., Rept. 1820). See Entheus and Phareas.

822. PELIA.*

1848. Doubl., Gen. Diurn. Lep., pl. 30: *Lamis*. Sole species, and therefore type.
1849. Ib., ib. 229: the same.
Subsequently used by Felder, but the name is preoccupied in Crustacea (Bell, 1835). See Peria.

823. PELION.*

1858. Kirb., List Brit. Rhop.: *Thaumas* (linea). Sole species, and therefore type.
The name is preoccupied; see the preceding. See also Adopæa.

824. PELLICIA.

1870. Plötz in Herr.-Schaeff., Correspondenzbl. Zoöl.-min. Ver. Regensb. xxiv. 159: Macareus, Macarius, chloracephala (chlorocephala), **dimidiatus** (dimidiata), and several MS. species.
Dimidiatus may be taken as the type.

825. PENETES.

1849. [Boisd. in] Doubl., Gen. Diurn. Lep., pl. 58: **Pamphanis**. Sole species, and therefore type.
1851. Boisd. in Westw., Gen. Diurn. Lep. 347: the same.
So used by Herrich-Schaeffer and Kirby.

826. PENTHEMA.

1848. Doubl., Gen. Diurn. Lep., pl. 39: **Lisarda**. Sole species, and therefore type.
1850. Westw., Gen. Diurn. Lep. 281: the same.
Subsequently used by Felder and Kirby. The name is unpleasantly near to Penthimia (Germ., Hemipt. 1821).

827. PENTILA.

1847. [Boisd. in] Doubl., List Br. Mus. 57 : **undularis**, and an unnamed species ;* but undularis was undescribed until 1866, by Hewitson.
1851? Westw., Gen. Diurn. Lep., pl. 76: Zymna.
1852. Boisd. in Westw., Gen. Diurn. Lep. 503: undularis [still inedited], Abraxas, Acræa, and, with a query, Evander.

Abraxas and Acræa are figured (pl. 77), but as species of Liptena, which Westwood at the time of the publication of the plates considered synonymous with Pentila. Zymna is placed in Miletus.

1866. Hewits., Exot. Butt. iii. 119: Mr. W. F. Kirby has kindly given me the following abstract of Hewitson's remarks, to which I have no personal access: —

"Westwood in error used Liptena on pl. 77 (Gen. Diurn. Lep.), and then adopted the name Pentila, applied by Boisduval to a part only, not knowing that Tingra was synonymous. Tingra was earlier, but uncharacterized, so Pentila must stand. Westwood's first species, undularis, was Boisduval's type of Pentila ; but, as Westwood's dissections were made from tropicalis,† it should stand as Westwood's type. Pentila includes tropicalis and Peucetia."

1868. Herr.-Schaeff., Prodr. iii. 13: uses it in the manner indicated by Hewitson.
1871. Kirb., Syn. Cat. 335: the same.

As will be seen from the above, the generic name, when first proposed, was founded upon undescribed species, and the characters of the genus were also withheld. It was first recognizable when Westwood figured a species, Zymna, under that name; but shortly afterward, and before any further use of the generic name, he confessed himself in error in supposing this to belong to Boisduval's genus. The latter he now characterized (probably from an examination of the first three species on his list), and placed therein the species first referred to it by Boisduval (though still unpublished), together with others, one of them doubtless the unnamed species referred to the genus by Doubleday. Still later, in 1866, Hewitson described undularis; and since, according to Kirby, it is congeneric with Acræa, it may most properly be considered the type of this genus, which should date from 1852. See Liptena and Tingra.

828. PEPLIA.

1816. Hübn., Verz. 20: Lamis, **caricæ**, Pelops (Pelope), Lysimon

---

* Probably Abraxas or Acræa.

† This cannot be true, for Westwood remarks of tropicalis: "An insect which I have not had an opportunity of examining," l. c. 504.

(Lisimæna), Molpe, Cachnus (Damæna), Menalcus (Menalcis), Mantus (Mante), Hebrus (Pelidna), Aristus (Ariste).

The group is synonymous with Desmozona and Heliochlæna, which fall before it. Caricæ may be taken as the type. See also Nymphidium.

829. Pepliphorus.

1816. Hübn., Verz. 71: Euchylas, **Cyanea** (Cyanus).

Cyanea may be taken as the type.

830. Pereute.

1867. Herr.-Schaeff., Prodr. ii. 11: **Callinice**, Charops (marina), Autodyca (Autodyce), Telthusa, Leucodrosime (Leucodrosyne).
1870. Butl., Cist. Ent. i. 34, 40: specifies Callinice as type.
1871. Kirb., Syn. Cat. 428: follows Herrich-Schaeffer, but includes in it the genus Leodonta.

831. Peria.

1871. Kirb., Syn. Cat. 205: **Lamis**. Sole species, and therefore type.

Proposed to supplant Pelia, of which Lamis was the type.

832. Perichares.

1872. Scudd., Syst. Rev. 60: **Corydon**, Trinitad, marmorata, Sandarac. Corydon is specified as type.

833. Peridromia.*

1836. Boisd., Spec. gén., pl. 7 C.: *Arethusa*. Sole species, and therefore type.

Used similarly by Douhleday, Felder, and Herrich-Schaeffer. The name is preoccupied through Peridroma (Hübn., Lep. 1816), which has the same derivation. See Ageronia.

834. Periplacis.

1837. Gey. in Hübn., Zutr. v. 32: **Glaucoma**. Sole species, and therefore type.

835. Periplysia.

1871. Gerst., Arch. f. Nat. xxxvii. i. 358: **Leda**. Sole species, and therefore type.
1873. Ib., Faun. Sans. 370: the same.

836. PERISAMA.

1849. Doubl., Gen. Diurn. Lep. 240: **Bonplandii**, Lebasii, D'Orbignyi (D'Orbignii), Euriclea, Humboldtii, Oppelii, Philinus? and a MS. species.
1861. Feld., Neues Lep. 20: no species are cited.
1871. Kirb., Syn. Cat. 208: follows Doubleday and adds other species.

Bonplandii may be taken as the type.

837. PEROPHTHALMA.

1851. Westw., Gen. Diurn. Lep. 455: **tenera**. Sole species, and therefore type.

838. PERRHYBRIS.

1816. Hübn., Verz. 91: **Pyrrha** (Eueidias), Medusa (Epimedusa).
1867. Herr.-Schaeff., Prodr. ii. 10: employs it for a large number of species, including Pyrrha, but not Medusa. Pyrrha is therefore the type.
1871. Kirb., Syn. Cat. 478: follows Herrich-Schaeffer.

839. PETAVIA.*

1828. Horsf., Descr. Cat. Lep. E. Ind. Co. 59, expl. pl. 2: *Petavius* (Sakuni).

This name falls, because derived from the species on which the genus is grounded. Moreover, we have nothing to do with it here, as it is not a butterfly.

840. PETREUS.*

1832-33. Swains., Zoöl. Ill. ii. 110: *Peleus* (Thetys). Sole species, and therefore type.

The plate consists solely of the earlier stages of a butterfly, belonging to the Tribuni. The characters of the group are drawn up partly from the perfect insect (represented on pl. 59, and referred to Marius, while here to the subgenus Petreus, — a nymphalideous insect) and half from the caterpillar, belonging, as stated, to a totally different group! The name therefore must be dropped altogether. Moreover, Petreus is one of the synonymes of Peleus. See also Athena.

841. PHÆDRA.*

1829. Horsf., Descr. Cat. Lep. E. Ind. Co. 123: *Thetys* (terricola, insularis). Sole species, and therefore type.
1868. Herr.-Schaeff., Prodr. iii. 19: employs it for a large number of species, including Thetys.

The name is taken from one of the synonymes of the species upon which the genus is grounded, and therefore it must be dropped. See Curetis, Anops, and Candalides.

842. Phædyma.

1861. Feld., Neues Lep. 31: **Heliodora**, Sankara.
Heliodora may be taken as the type.

843. Phalanta.*

1829. Horsf., Descr. Cat. Lep. E. Ind. Co. expl. pl. 7: *Phalanta*.
Sole species, and therefore type.
As the name is founded upon that of the only species, it falls. See Atella.

844. Pnanessa.*

1837. Sodoffsk., Bull. Mosc. x. 80: proposed as a more correct spelling for Vanessa.

845. Pnanus.

1816. Hübn., Verz. 114: **vitreus.** Sole species, and therefore type.
Subsequently thus used by Butler and Herrich-Schaeffer.

846. Phareas.*

1852. Westw., Gen. Diurn. Lep. 515: Eumelus, Dumerilii, Talaus, Busiris, Peleus, Gentius, Procas, Pertinax, Cœleste, Loxus, Tertullianus, Julettus. Gentius and Peleus specified as typical.

1869. Butl., Cat. Fabr. Lep. 283: employs it for six species, including Gentius.
Peleus being already type of Entheus, and Gentius being strictly congeneric, this name must give place to Entheus. See also Peleus.

847. Phasis:

1816. Hübn., Verz. 73: **Thero** (Salmoneus), Thysbe (Palmus, Nais).

1869. Butl., Cat. Fabr. Lep. 176: employs it for Thero (Rumina), which thereby becomes type.

848. Pheles.

1858. Boisd. in Herr.-Schaeff., Exot. Schmett. 77: **Heliconides.**
Sole species, and therefore type.
Used in same sense by Bates and Kirby.

849. Phemiades.

1816. Hübn., Verz. 112: Ephesus, Edipus (Edippus), Epictetus, **Phineus,** Augias.
Phineus may be taken as the type.

850. PHILÆTHRIA.

1820. Dalm. in Billb., Enum. Ins. 77: I. Hippona; II. **Dido**, Phærusa, Julia.

Dido may be taken as the type.

851. PHILOCALA.

1820. Billb., Enum. Ins. 79: Feronia, **Amphinome**, Orithya, Genoveva, Œnone, cardui, Huntera, Atalanta (Atalantha), Levana (Prorsa, Levana), Polynice, Ilithya.

Felder (Neues Lep. 17) divides Ageronia into four sections, the second of which, unnamed, contains only Amphinome. This may be taken as the type of Philocala.

852. PHILOGNOMA.*

1844. [Boisd. in] Doubl., List Br. Mus. 112: Decius, Varanes.
1850. Boisd. in Westw., Gen. Diurn. Lep. 310: Decius, Varanes, Laodice, Lichas. The latter is figured.

The name falls before Palla.

853. PHILONOMA.*

1820. Billb., Enum. Ins. 78: proposes, without reason, to employ this name for Neptis. Of course it falls.

854. PHILOODUS.

1840. Ramb., Faune Ent. Andal. ii. 308: **Nostrodamus** (Nostradamus, Lefebvrei). Sole species, and therefore type.

855. PHLEBODES.

1816. Hübn., Verz. 107: **Pertinax**, Saturnus.
1870. Butl., Ent. Monthl. Mag. vii. 93: Pertinax is specified as the type.

856. PHLOGRIS.*

1822-26. Hübn., Exot. Schmett. ii.: *Melpomene*. Sole species, and therefore type.

The name falls before Sunias (q. v.).

857. PHOCIDES.

1816. Hübn., Verz. 103: **Palemon** (cruentus), Alardus, Lisias (Lisiades).

Lisias does not belong to the family in which this was placed by Hübner, and therefore cannot be taken as the type, Alardus belongs to Telegonus, and therefore Palemon must be taken as the type. See Dyscnius.

858. Phœbis.

1816. Hübn., Verz. 98: Philea (Melanippe), Crocale (Jugurtha), **Argante** (Cypris), Eubule (Eubule, Drya).

1873. Butl., Lep. Exot. i. 155: designates Argante (Cypris) as the type.

859. Pholisora.

1872. Scudd., Syst. Rev. 51: **Catullus**, Hayhurstii, Azteca. Catullus specified as type.

860. Phorcis.*

1816. Hübn., Verz. 62: Scœa, Stygne (Epistygne), Gorge.

The name falls before Erebia. See also Gorgo, Marica, Syngea, Epigea, and Oreina.

861. Phrissura.

1870. Butl.,' Cist. Ent. i. 37, 49: Cynis. Sole species and designated type. [See, however, additional note, p. 293.]

862. Phryne.*

1843. Herr.-Schaeff., Schmett. Eur. i. 90: *Phryne* (Tircis). Sole species, and therefore type.

The name is taken from one of the synonymes of the species upon which the genus was founded, and therefore falls. It may also be pre-occupied in Reptiles (Fitz. 1843). See Triphysa.

863. Phulia.

1867. Herr.-Schaeff., Prodr. ii. 17: **Nymphula**. Sole species, and therefore type, as stated by Butler.

864. Phycanassa.

1872. Scudd., Syst. Rev. 56: **Viator**. Sole species and designated type.

865. Phyciodes.

1816. Hübn., Verz. 29: **Tharos** (Cocyta), Liriope.

1848. Doubl., Gen. Diurn. Lep. 181: employs it for these and many others.

1850. Steph., Cat. Brit. Lep. 259: uses it for Tharos only, which thereby becomes type.

1872. Scudd., Syst. Rev. 25: specifies Tharos (Cocyta) as type.

866. Phyllophasis.*

1841. Blanch., Hist. Ins. iii. 447: Galanthis (Galanthus), Varanes (Veranes).

This name falls before Palla and Siderone.

867. PHYSCÆNEURA.

1857. Wallengr., Rhop. Caffr. 32: **Panda**. Sole species, and therefore type.

868. PHYTALA.

1847. [Boisd. in] Doubl., List Br. Mus. 20: **Elais**. Sole species, but unpublished, and the genus uncharacterized.
1852. Boisd. in Westw., Gen. Diurn. Lep. 471: **Elais**. The species figured.
   It is therefore type, and the genus should bear date 1852.

869. PICANOPTERYX.

1857. Wallengr., Rhop. Caffr. 7: I. Severina, Gidica (Doubledayi), Mesentina, Gidica (Westwoodi); II. **Eriphia**, Charina (Simana, alba).
1858. Ib., K. Vetensk. Acad. Förh. xv. 75: I. Severina; II. Eriphia, Ada.
   The first section being synonymous with Belenois, Eriphia may be taken as the type. See Herpænia.

870. PIERELLA.

1851. Westw., Gen. Diurn. Lep. 365: **Nereis**, Rhea (Dindymene), Lena, Astyoche (Larymna), Dracontis (Lena, Dracontis).
1864. Herr.-Schaeff., Prodr. i. 55: the same and others.
1868. Butl., Ent. Monthl. Mag. iv. 195; Cat. Sat. 103: specifies Nereis as type.

871. PIERIS.

1801. Schrank, Fauna Boica, ii. i. 152, 160: I. Apollo; II. Polyxena, Machaon, Podalirius; III. cratægi, brassicæ, rapæ, napi, sinapis, Daplidice, cardamines; IV. rhamni, Palæno, Hyale.
1805. Latr., Sonn. Buff. xiv. 111: restricts it to the third and fourth groups, specifying, of Schrank's species, rhamni, Hyale, cratægi, brassicæ, napi, Daplidice, sinapis, cardamines.
1809. Ib., Gen. Crust. et Ins. iv. 203: restricts it still further to Schrank's third section, and divides it thus: I. *a*, cratægi, brassicæ, Daplidice, cardamines, etc.; *b*, sinapis; II. Crisia.
1810. Ib., Consid. 440: specifies brassicæ as type; but that is already type of Mancipium (q. v.).
1815. Oken, Lehrb. i. 727: employs it for the swallow-tails.

1816. Lam., Hist. Nat. An. sans Vert. iv. 30: brassicæ and others, including rapæ.
1816. Hübn., Verz. 53: employs it for species of another family.
1827. Steph., Ill. Brit. Ent. Haust. i. 25: restricts it to cratægi, but improperly. See Aporia.
1831. Curtis, Brit. Ent., pl. 360: also specifies cratægi as type.
1832-33. Swains., Zoöl. Ill. ii. 69: designates Belisama as type; but it is not even one of Schrank's species.
1836. Boisd., Spec. gén. 434: employs it for all of Schrank's third group, excepting sinapis and cardamines, and this has been its general use ever since.

Since all of the other species given by Schrank must be taken as types of other genera (see Aporia, Mancipium, Pontia, Euchloe, and Leptidia), rapæ may be taken as the type, as it is virtually included in the group by Latreille in 1809. This, too, would best accord with modern usage.

1840. Westw., Gen. Syn. 87: also specifies cratægi as type.
1870. Butl., Cist. Ent. i. 37, 49: considers Demophile (Amathonte) as type. This, he says, is the type of Boisduval's Pieris [it was simply his first species]; Apollo, that of Schrank's; Leucippe, that of Latreille and Godart; he adopts Demophile, because "we ought to have a genus Pieris in the Pierinæ." But it was not mentioned by Schrank.
1872. Scudd., Syst. Rev. 41: specifies brassicæ as type, erroneously. See Mancipium, Ganoris, and Catophaga.

872. PIERITES* (fossil).

1849. Herr, Insektenf. Oening. ii. 182: *Freyeri.* Sole species, and therefore type.

Falls before Synchloe, with which it is synonymous. In 1840 a group of butterflies was named Piérites by Blanchard and Brullé; but it would hardly affect this, for the name is not Latin, but a Gallicized form of Latin.

873. PINDIS.

1869. Feld., Verhandl. Zoöl.-bot. Gesellsch. Wien. xix. 475: **squamistriga.** Sole species, and therefore type.
1871. Kirb., Syn. Cat. 108: the same.

874. PISOLA.

1865. Moore, Proc. Zoöl. Soc. Lond. 785: **Zennara.** Sole species, and therefore type.
1871. Kirb., Syn. Cat. 583: the same.

875. PITHECOPS.

1828. Horsf., Descr. Cat. Lep. E. Ind. Co. 66: **Hylax.** Sole species, and therefore type.
1847. Doubl., List Br. Mus. 57: the same.
1850. Steph., Cat. Brit. Lep. 18: employs it for Argiolus.

876. PLANEMA.

1848. Doubl., Gen. Diurn. Lep. 140: I. **Lycoa,** Gea (Jodutta, Carmentis); II. Gea, Euryta (Euryta, Umbra).
 Lycoa may be taken as the type.

877. PLASTINGIA.

1870. Butl., Ent. Monthl. Mag. vii. 95: **flavescens,** tessellata, Callineura, extrusa. Flavescens specified as type.
1871. Kirb., Syn. Cat. 619: the same and others.

878. PLEBEIUS.*

1871. Kirb., Syn. Cat. App. 653: proposes to use this in the place of Cupido (in which he places most of the blues), as having been founded by Linné. Three hundred and twenty-nine species are included in it.
1872. Crotch, Cist. Ent. i. 60: says, wrongly, that Linné used this name in a generic sense, and that Cuvier, in 1799, fixed its type as Argus. See Introductory Remarks.

879. PLESIONEURA.*

1862. Feld., Wien. Ent. Monatschr. vi. 29: *curvifascia.* Sole species, and therefore type, as stated by Butler.
1871. Kirb., Syn. Cat. 620: employs it for this and others.
 The name is preoccupied in Diptera (Macq. 1855). See Celænorhinus.

880. POANES.

1872. Scudd., Syst. Rev. 55: **Massasoit.** Sole species and designated type.

881. PODALIRIUS.*

1832–33. Swains., Zoöl. Ill. ii. 105: Antiphates (Pompilius), *Podalirius* (Europæus). Podalirius specified as type.
 The name being founded upon that of one of its species, it falls. It is also preoccupied in Hymenoptera (Latr. 1802). See Iphiclides.

882. POLITES.

1872. Scudd., Syst. Rev. 57: **Peckius,** Sabuleti. Peckius specified as type.

883. POLYCHROA.

1820. Billb., Enum. Ins. 78: **Obrinus**, Ancæus.
Obrinus may be taken as the type.

884. POLYCYMA.*

1862. Scott in Feld., Verh. Zoöl.-bot. Gesellsch. Wien. xii. 490: Felder says that Scott [in litt.?] proposes this for the species, which Felder there places in Holochila. Felder does not adopt the name, because it is not appropriate for most of the species. See also Erina.

885. POLYDORUS.*

1832–33. Swains., Zoöl. Ill. ii. 101: Aristolochiæ (Thoas), Polydorus, Polytes (Polystes, Romulus). The last two specified as types.
As the name is founded upon that of one of the species included in it, it falls. Moreover, it is preoccupied through Polydora (Bosc, Worms, 1802). See Pachlioptera.

886. POLYGONIA.

1816. Hübn., Verz. 36: Egea (triangulum, i. album), c. aureum, Progne, **c. album**.
1858. Kirb., List Brit. Rhop.: employs it for c. album only, which therefore becomes type.
1872. Scudd., Syst. Rev. 9: specifies c. aureum as type, but incorrectly.
See Grapta and Comma.

887. POLYGONUS.*

1822–26. Hübn., Exot. Schmett. ii.: *Amyntas* (lividus). Sole species, and therefore type.
The name is preoccupied through Polygona (Schum., Moll. 1817), and is very close to Hübner's own Polygonia. See Acolastus.

888. POLYOMMATUS.

1805. Latr., Sonn. Buff. xiv. 116: betulæ, quercus, pruni, **Bœticus**, rubi, Argus, Thetis (Adonis), Endymion (Meleager), Corydon, Arion, Arcas (Erebus), Cyllarus, Semiargus (Acis), Argiolus, Alsus, Dorilas (Myopa), Phlæas, virgaureæ. Corydon alone is figured.
1807. Ib., Gen. Crust. et Ins. iv. 206: divides the group into sections, specifying a few, as follows: I. *a*, betulæ, quercus, and others not in previous list; *b*, Bœticus, Endymion (Meleager), rubi, Phlæas, virgaureæ; II. Argus, Corydon, Alsus.

1810. Ib., Consid. 440: specifies betulæ, quercus, Bœticus, and Argus as types.
1817. Ib. in Cuv., Règne An. iii. 553: specifies only Alexis (not given in the original list), as a species found in the environs of Paris, but refers to previous works for the species.
1823. God., Encycl. méth. 595: employs it for all Ephori, including all the above species.
1823. Ib., Tab. Méth. 46: does the same.
1828. Horsf., Descr. Cat. Lep. E. Ind. Co. 67: restricts the name to the blues, but only uses it for new species.
1828. Steph., Ill. Brit. Ent. Haust. 83: uses it for Argiolus and many others, all blues, including Argus.
1829. Boisd., Index, 10: follows Godart.
1830. Meig., Eur. Schmett. ii. 1: employs it for the same, excluding the hair-streaks.
1832. Renn., Consp. 17: uses it for the blues only, specifying, of those given by Latreille in the first instance, Argus, Thetis (Adonis), Corydon, Arion, Semiargus (Acis), Argiolus, Alsus, and Dorylas.
1832–33. Swains., Zoöl. Ill. ii. 133: uses it for Cassius, one of the blues.
1832–33. Boisd., Icones, 43: employs it for the coppers only.
1832. Dup., Pap. France, Diurn. Suppl. 391: the same.
1833–34. Boisd.-LeC., Lép. Amer. Sept. 122: the same.
1839. Ramb., Faune Ent. Andal. 264: places the blues here again.
1840. Westw., Gen. Syn. 88: specifies Arion as type.
1847. Doubl., List Br. Mus. 53: follows Boisduval, as do most subsequent authors.
1870. Kirb., Journ. Linn. Soc. Lond. x. 500: thinks that Corydon should be taken as the type, because figured in the first instance by Latreille.

Latreille's own action necessitates its restriction to the blues; but Corydon cannot be taken as the type, since it belongs to Rusticus, established in 1810. Nor can Argus for the same reason. The only other type of blues mentioned by him in 1810 is Bœticus, for which Polyommatus must be retained. See also Lycæna.

889. POLYSTICHTIS.

1816. Hübn., Verz. 18: Fatima (Cerca), **Zeanger** (Zeangira), Mandana (Mandane), Lucinda.

1872. Scudd., Syst. Rev. 28: specifies Cæneus as type.

Erroneously, through Hübner's confounding of that species with the Fatima of Cramer. Hübner's first species (No. 109) must be referred primarily to Cramer's Fatima, because he appends a mark of exclamation or approval, after the reference to his figs. A. B., and of interrogation or doubt to his C. D., showing that Cramer's A. B. (Fatima) was in Hübner's mind, unquestionably, the species referred to by his No. 109.

1873. Grote, Can. Ent. v. 144: corrects the identification of Scudder, and suggests that Fatima should be taken as the type.

This, however, became in 1818 the type of Emesis; so, too, Lucinda was placed, by another name, under Aphacitis, and must be taken as the type of that genus. Mandana belongs to Emesis, and consequently Zeanger must be taken as the type, and Polystichtis may replace Lemonias auct. nec Hübn. (Tent.). See Calospila.

890. POLYURA.

1820. Billb., Enum. Ins. 79: Jason (Jasius), **Pyrrhus.**
Pyrrhus may be taken as the type.

891. PONTIA.

1807. Fabr., Ill. Mag. vi. 283: cratægi, rapæ, **Daplidice,** Elathea, bella.

1815. Leach, Edinb. Encycl. 716: cratægi, brassicæ, rapæ, napi, cardamines, Daplidice, sinapis.

1816. Ochs., Schmett. Eur. iv. 30: employs it for cratægi, rapæ, Daplidice, and others.

1816. Hübn., Verz. 92: uses it for Hyparete, Eucharia, and Hierte of the same family.

1824. Curtis, Brit. Ent. pl. 48: designates Daplidice as type, which must stand, although seldom used since in this manner.

1827. Steph., Ill. Brit. Ent. Haust. i. 14: uses it for rapæ and others not in Fabricius's list, placing cratægi and Daplidice elsewhere; thus indicating rapæ as the type.

1829. Horsf., Descr. Cat. Lep. E. Ind. Co. 138, 142: divides it into several named groups, and places in Pontia proper a number of species distantly allied to those of Fabricius.

1836. Boisd., Spec. gén. 430: restricts it to several species of whites not mentioned by Fabricius.

1840. Westw., Gen. Syn. 87: specifies brassicæ as the type.

1844. Doubl., List Br. Mus. 24: follows Boisduval.

1847. Ib., Gen. Diurn. Lep. 40: does the same.

1867. Herr.-Schæff., Prodr. ii. 8: the same.
1870. Butl., Cist. Ent. i. 38, 50: designates cratægi as the type.
1871. Kirb., Syn. Cat. 439: follows Boisduval.
1872. Crotch, Cist. Ent. i. 66: designates Xiphia (Nina) as type, through Boisduval in 1836.
 See Ganoris, Mancipium, and Synchloe.

892. PORITIA.*

1865. Moore, Proc. Zoöl. Soc. Lond. 775: *Hewitsoni.* Sole species, and therefore type.
1871. Kirb., Syn. Cat. 409: the same.
 The name is, correctly speaking, preoccupied, through Porites (Lam., Pol. 1816).

893. POTAMIS.

1806. Hübn., Tent. 1: **Iris.** Sole species, and therefore type.
 This name, never since used, must be restored. See Apatura.

894. POTANTHUS.

1872. Scudd., Syst. Rev. 54: **Omaha,** Californica. Omaha specified as type.

895. PRECIS.

1816. Hübn., Verz. 33: **Octavia,** Dryope.
1849. Doubl., Gen. Diurn. Lep. 209: employs it for Octavia and others, to the exclusion of Dryope; and the former therefore becomes the type.
 It has since been used in the same sense by Felder, Butler, and Kirby.

896. PRENES.

1872. Scudd., Syst. Rev. 60: **Panoquin,** Ocola, Hecebolus, sylvicola. Panoquin specified as type.

897. PREPONA.

1836. Boisd., Spec. gén., pl. 3 B.: **Laertes** (Demodice). Sole species, and therefore type.
 Since used in same sense by Doubleday, Westwood, Felder, and Kirby.

898. PRIAMIDES.

1816. Hübn., Verz. 87: Torquatus (Caudius), **Pompeius** (Hipponous, Capys), Echelus (Echemon), Furisteus, Æneas (Marcius), Sesostris (Tullus), Anchises (Anchises, Brissonius, Pompejus), Hippason (Amosis, Hippason).
 Pompeius may be taken as the type.

899. PRINCEPS.

1806. Hübn., Tent. 1: **Machaon**. Sole species, and therefore type.
See Amaryssus, Papilio.

900. PRIONERIS.

1867. Wall., Trans. Ent. Soc. Lond. [3] iv. 383: **Thestylis** (Thestylis, Seta), Sita, Clemanthe (Clemanthe, Berenice), Vollenhovii, Cornelia, Philonome, Autothisbe.
1870. Butl., Cist. Ent. i. 33: specifies Thestylis as the type.
1871. Kirb., Syn. Cat. 477: employs it for all of Wallace's species and others.

901. PROCRIS.

1864. Herr.-Schaeff., Prodr. i. 23: no species mentioned.

In his list, p. 66, this name is supplanted by Acca Hübn. and Procris and Urdaneta referred to it. These cannot be placed in Acca (q.v.); but the name of the genus, being the same as that of one of the species upon which it is founded, falls. It is also preoccupied in Lepidoptera (Fabr., 1807).

902. PROMETHEUS.*

1822-26. Hübn., Exot. Schmett. ii.: *Casmilus*. Sole species, and therefore type.

It is not a butterfly.

903. PRONOPHILA.

1849. Doubl., Gen. Diurn. Lep., pl. 60: **Thelebe**, Irmina, Phoronea.
1850. Doubl., Gen. Diurn. Lep., pl. 66: Tauropolis.
1851. Westw., Gen. Diurn. Lep. 357: the same with others.
1867. Butl., Ann. Mag. Nat. Hist. [3] xx. 266; Cat. Sat. 184: specifies Thelebe as the type.
1871. Kirb., Syn. Cat. 108: uses it in Butler's sense.

904. PROTEIDES.

1816. Hübn., Verz. 105: **Idas** (Mercurius), Zestos, Exadeus, Lycidas (Lyciades), Clonius, Renaldus, Assaricus, Amphion.
1869. Butl., Cat. Fabr. Lep. 264: employs it for seven species, including Idas and Clonius (Clonias).
1870. Ib., Ent. Monthl. Mag. vii. 93: specifies Idas (Mercurius) as type.

905. PROTESILAUS.*

1832-33. Swains., Zoöl. Ill. ii. 93, 104: *Protesilaus* (Leilus), Bellerophon (Swainsonii). Protesilaus specified as type.

The name, of course, falls, from being founded upon one of the species on which the genus is established.

906. PROTHOE.

1822–26. Hübn., Exot. Schmett. ii.: **Franckii.** Sole species, and therefore type. See Œnomaus.

Used in same sense by Doubleday, Felder, and Kirby.

1850. Westw., Gen. Diurn. Lep. 266: employs it for this species only, and gives Autonema as a MS. generic synonyme of Boisduval.

907. PROTOGONIOMORPHA.

1857. Wallengr., Rhop. Caffr. 23: **Anacardii.** Sole species, and therefore type.

1861. Feld., Neues Lep. 14: Sabina, Anacardii.

Should Anacardii prove congeneric with Augustina (as given by Kirby), this name will fall before Salamis.

908. PROTOGONIUS.*

1816. Hübn., Verz. 100: *Hippona* (Fabius). Sole species, and therefore type.

Used for this species only, by Westwood, Felder, Butler, and Kirby; but the name falls before Consul. See also Fabius and Helicodes.

909. PSALIDOPTERIS.

1822. Hübn., Zutr. ii. 17: **Thucydides** (Nycha). Sole species, and therefore type.

1837. Gey. in Ib., v. 26: Terambus (Lytæa). A very different insect. See Theope.

910. PSELNA.*

1820. Billb., Enum. Ins. 77: proposes, without reason, to use this name for Hætera (q. v.).

911. PSEUDACRÆA.

1850. Westw., Gen. Diurn. Lep. 281: **Hirce**, Euryta, Boisduvalii.

1871. Kirb., Syn. Cat. 229: employs it for Hirce, Boisduvalii, and others.

Hirce may be taken as the type. See Panopea.

912. PSEUDERGOLIS.

1867. Feld., Reise Novara, 404: **Avesta.** Sole species, and therefore type.

913. PSEUDODIPSAS.

1860. Feld., Wien. Ent. Monatschr. iv. 243: **Eone.** Sole species, and therefore type.

1871. Kirb., Syn. Cat. 408: the same and others.

914. PSEUDOLYCÆNA.

1858. Wallengr., K. Vet. Akad. Förh. xv. 80: **Marsyas**. Sole species, and therefore type. See Œnomaus.

915. PSEUDONYMPHA.*

1857. Wallengr., Rhop. Caffr. 31: *Hippia*, Cassius (hyperbioides), Hyperbius, Narycia.
1858. Ib., K. Vet. Akad. Förh. xv. 79: Hippia only, which therefore becomes type.
1868. Butl., Ent. Monthl. Mag. iv. 194; Cat. Sat. 93: specifies Hippia as the type.

This name must fall before Melampias.

916. PSEUDOPHELES.*

1867. Bates, Trans. Ent. Soc. Lond. [3] v. 544: *Sericina*. Sole species, and therefore type.

The name falls before Esthemopsis, as pointed out by Bates himself.

917. PSEUDOPONTIA.*

1870 (Sept.). Plötz, Stett. Ent. Zeit. xxxi. 348: *paradoxa* (calabarica). Sole species, and therefore type.
1870. Butl., Cist. Ent. i. 57: expresses the opinion that it is not a butterfly, but a moth.
1871. Kirb., Syn. Cat. 438: employs it for the same.

The name falls before Gonophlebia. See also Globiceps.

918. PTERONYMIA.

1872. Butl.-Druce, Cist. Ent. i. 96: **Aletta**, Olyrilla, Notilla, fulvimargo. Aetta specified as type.

919. PTEROURUS.*

1777. Scop., Introd. 433: Paris and a great number of others destitute of the slightest distinguishing bond of union of any value.

They are mostly butterflies whose hind wings are prolonged into a tail. They are divided into two sections, but it would be difficult to say on what ground. The second section contains the following, among others: Hector [Papilionides], Leilus [Urania], pruni [Ephori], Proteus [Urbicolæ], Butes [Vestales].

1872. Scudd., Syst. Rev. 43: specifies Troilus (one of the Scopolian species) as type. See Euphœades.

But unreasonably and indefensibly, as the name must fall from the incongruity of the materials of which the genus is composed.

920. PTERYGOSPIDEA.

1857. Wallengr., Rhop. Caffr. 53: **Flesus** (Ophion), Motozi, Mokeesi, Sabadius (Nottoana).
1858. Ib., K. Vet. Akad. Förh. xv. 83: Flesus (Ophion) and a new species. Flesus therefore becomes type.

921. PTYCHANDRA.

1861. Feld., Wien. Ent. Monatschr. v. 304: **Lorquinii**. Sole species, and therefore type, as stated by Butler and used by different authors.

922. PTYCHOPTERYX.*

1857. Wallengr., Rhop. Caffr. 17: *subfasciatus* (Bohemanni). Sole species, and therefore type, as stated by Butler.

The name, however, is preoccupied in Diptera (Leach, 1818), and it was probably on this account that Wallengren subsequently proposed Thespia (q. v.) in its stead. The name falls before Teracolus.

923. PYCINA.

1849. Boisd. in Doubl., Gen. Diurn. Lep., pl. 48: **Zamba**. Sole species, and therefore type.
1850. Boisd. in Westw., ib. 305: the same.

Subsequently used by Felder and Kirby.

924. PYRAMEIS.*

1816. Hübn., Verz. 33: Indica (Calliroe), *Atalanta*.
1849. Doubl., Gen. Diurn. Lep. 202: employs the name for these and others, placing them in two sections, both of Hübner's in the first.
1850. Steph., Cat. Brit. Lep. 11: uses it for Atalanta only, which therefore becomes the type.

But Atalanta is already the type of Vanessa, and both species are strictly congeneric; consequently this name falls. See also Ammiralis and Bassaris.

925. PYRGUS.*

1816. Hübn., Verz. 109: *Syrichtus* (Syrichtus, Oilus, Orcus), sidæ, Tessellum, Alveus (carthami), Fritillum, malvæ (Alveolus), Sao (Sertorius), Vindex.
1850. Steph., Cat. Brit. Lep. 21, 262: employs it for malvæ (Alveolus), Syrichtus (Oileus), and alceæ (malvarum).
1852. Westw., Gen. Diurn. Lep. 516: uses it for all of Hübner's species and for others.

1858. Kirb., List Brit. Rhop.: follows Stephens.
1869. Butl., Cat. Fabr. Lep. 280: employs it for Syrichtus, sidæ, malvæ, and others.
1870. Ib., Ent. Monthl. Mag. vii. 94: specifies Syrichtus as type.

 The name falls before Hesperia, all the species being strictly congeneric with malvæ, the type of that genus. See also Scelothrix and Syrichtus.

926. Pyristia.

1870. Butl., Cist. Ent. i. 35, 44: **Proterpia.** Sole species and designated type.

927. Pyronia.

1816. Hübn., Verz. 59: **Tithonus** (Tithone), Ida, Narica.
1850. Steph., Cat. Brit. Lep. 7: employs it for Tithonus only.

 In this he is followed by Kirby (List, 1858), and this may be considered the type.

928. Pyrrhogyra.

1816. Hübn., Verz. 43: **Tipha**, Neærea.
1844. Doubl., List Br. Mus. 88 [Pyrrhagyra]: employs it for Tipha only, which thereby becomes type.
1850. Westw., Gen. Diurn. Lep. 252: employs it for both of Hübner's species and others. See also Corybas.

 Subsequently employed similarly by Felder and Kirby.

929. Pyrrhopyge.

1816. Hübn., Verz. 103: Phidias (Bixæ), **hyperici**, Acastus (Phidias), Amyclas, Arinas.
1852. Westw., Gen. Diurn. Lep. 508 [Pyrrhopyga]: employs it for fourteen species, including all but the last of Hübner's, and adding others.
1869. Herr.-Schaeff., Prodr. iii. 56 [Pyrrhopyga]: uses it for a still greater number of species, including all of Hübner's.
1869. Butl., Cat. Fabr. Lep. 267 [Pyrrhopyga]: refers to it all of Hübner's species excepting hyperici, and adds others.
1870. Ib., Ent. Monthl. Mag. vii. 58 [Pyrrhopyga]: places here all of Hübner's species excepting Arinas, and adds several others.
1871. Kirb., Syn. Cat. 584: employs it for all of Hübner's species and others.

1872. Scudd., Syst. Rev. 46 [Pyrrhopyga]: specifies Phidias (Bixæ) as type.

> Phidias, however, was taken in 1852 as type of Pachyrhopala. Hyperici may be selected as the type of this genus.

930. PYRRHOSTICTA.

1872. Butl., Cist. Ent. i. 86: Lætitia "and allies." Lætitia is then the type.

931. PYTHONIDES.

1816. Hübn., Verz. 111; Jovianus, **Cerialis** (Cerberus), Lagia (Herennius).
1827–37. Gey. in Hübn., Exot. Schmett. iii. [Pithonides]: employs it for Cerialis (Orcus) and Lagia (Herennius).
1869. Butl., Cat. Fabr. Lep. 285: uses it for Jovianus, Cerialis (Cerealis), and another.
1870. Ib., Ent. Monthl. Mag. vii. 97: specifies Jovianus as type.
1871. Kirb., Syn. Cat. 626: uses it for all these species and others.

> Jovianus, which is generically distinct from Cerialis, cannot be taken as the type, because left out of the group by Geyer. Cerealis may be taken as the only one used by all authors.

932. RAGADIA.

1851. Westw., Gen. Diurn. Lep. 376: **Crisia.** Sole species, and therefore type, as stated and employed by Butler.
1871. Kirb., Syn. Cat. 56: Crisia, Crisilda.

933. RHAPHICERA.

1867. Butl., Ann. Mag. Nat. Hist. [3] xix. 164: **Satricus,** Moorei.
1868. Ib., Ent. Monthl. Mag. iv. 196; Cat. Sat. 158: specifies Satricus as type.

934. RHETUS.*

1832–33. Swains., Zoöl. Ill. ii. 33: Butes (Crameri), Rhetus, Periander. The last two are specified as types.

> Afterward employed by Westwood (Gen. Diurn. Lep.), but the name must fall because based on that of one of the species upon which it was established. It is also preoccupied through Rhetia (Leach, Crust. 1818). See Diorina.

935. RHINOPALPA.*

1860. Feld., Wien. Ent. Monatschr. iv. 399: *fulva.* Sole species, and therefore type.
1861. Ib., Neues Lep. 49: Polynice, fulva.

1871. Kirb., Syn. Cat. 191: the same and others.

Mr. Kirby suggests to me that this name is probably hybrid (ῥίσ, palpus), and on that account changed by Felder himself to Eurhinia (q.v.), just as he changed Teinopalpus to Teinoprosopus.

936. RHODOCERA.

1829. Boisd.-LeC., 70: Mærula, rhamni, Clorinde, **Menippe** (Leachiana).
1832. Dup., Pap. France, Diurn. Suppl. 386: uses it for rhamni and Cleopatra.
1836. Boisd., Spec. gén. 597: employs it for the same species as Boisduval and LeConte, and for others.
1840. Ramb., Faune Ent. Andal. ii. 256: employs it for Cleopatra only.
1844. Doubl., List Br. Mus. 37: follows Boisduval's practice in 1836.
1847. Ib., Gen. Diurn. Lep. 70: suggests that it be used for the American species placed in that work under Gonepteryx, namely, Menippe (Leachiana), Clorinde, and Mærula of Boisduval and LeConte's list, and a few others. See also Amynthia.
1870. Butl., Cist. Ent. i. 35: specifies Menippe (Leachiana) as type.

It cannot be taken for rhamni and allies, as Duponchel's action would require, because they were reserved for Colias as early as 1810. We may therefore follow Doubleday (1847), through Butler, in considering Menippe as the type.

937. RHOPALOCAMPTA.

1857. Wallengr., Rhop. Caffr. 47: **Forestan** (Florestan), Valmaran, Keithloa.
1858. Ib., K. Vet. Akad. Förh. xv. 81: employs it for Forestan (Florestan) only, which thereby becomes the type.

938. RIODINA.

1851. Westw., Gen. Diurn. Lep. 430: **Lysippus.** Sole species, and therefore type.

Thus used, for this species only, by Bates and Kirby. See Erycina.

939. RIPHEUS.*

1832–33. Swains., Zoöl. Ill. ii. 131: *Dasycephalus.* Sole species, and therefore type.

The name will fall because derived from a species of Drury's (Ripheus), with which this is directly compared. Moreover, it is probably a fictitious insect, having the appearance of a Uranian to which clubbed antennæ have been artificially attached.

940. RODINIA.*

1851. Westw., Gen. Diurn. Lep. 430: Jurgensenii (Jurgensenii, Montezuma), Calphurnia (Calpharnia), Periander, Aulestes (Aulestes, Glaphyra), Pandama, Tedea, Melibœus, (Melibœus, Julia), Inca.

<small>The name must fall, because the species mentioned belong to the earlier genera Ancyluris, Diorina, Zeonia, and Eueryoina.</small>

941. ROMALEOSOMA.

1840. Blanch., Hist. Ins. iii. 448: **Eleus.** Sole species, and therefore type.
1844. Doubl., List Br. Mus. 99: Eleus and others.
1850. Westw., Gen. Diurn. Lep. 283 [Romalæosoma]: the same in three sections.

942. RUSALKIA.

1871. Kirb., Syn. Cat. 306: **Marathon,** Ctesiphon.
1873. Ib., Zoöl. Rec. for 1871, 364: Marathon given as type. See Oreas.

943. RUSTICUS.

1806. Hübn., Tent. 1: **Argyrognomon** (Argus). Sole species, and therefore type.

<small>See Lycæides, Scolitantides, and Polyommatus.</small>

944. SAGARITIS.

1822–26. Hübn., Exot. Schmett. ii: **Orsis** (Orseis). Sole species, and therefore type. See Myscelia.

945. SAIS.

1816. Hübn., Verz. 10: **Rosalia,** Pyrrha (Pamela).
1844. Doubl., List Br. Mus. 57: employs it for Rosalia and some unnamed species.
1848. Ib., Gen. Diurn. Lep. 131: uses it for Rosalia and Cyrianassa.
1862. Bates, Linn. Trans. xxiii. 527: specifies Rosalia as the type.

946. SALACIA.*

1823. Hübn., Zutr. ii. 25: *Phyllodoce.* Sole species, and therefore type.

<small>The name, however, is preoccupied in Polyps (Lamx. 1816). See Scada.</small>

947. SALAMIS.

1833. Boisd., Ann. Mus. Hist. Nat. 194: **Augustina.** Sole species, and therefore type.
1844. Doubl., List Br. Mus. 84: employs it for a large number of species, not including Augustina.

1849. Ib., Gen. Diurn. Lep. 211: restricts it to half a dozen species, including Augustina.
1861. Feld., Fam. Nymph. 13: divides it into two sections, but does not specify Augustina in either.
1871. Kirb., Syn. Cat. 192: follows Doubleday (1849).
See Protogoniomorpha.

948. SALPINX.

1816. Hübn., Verz. 17: **leucostictos** (Nemertes). Sole species, and therefore type.

949. SAROTA.

1851. Westw., Gen. Diurn. Lep. 424: Dematria, **Chrysus**.
Chrysus may be taken as the type.

950. SARROMIA.*

1851. Westw., Gen. Diurn. Lep., pl. 67: *obsoleta*. Sole species, and therefore type.
This name falls before Lymanopoda, proposed at the same time, but subsequently united by their author under the latter name (q. v.).

951. SATARUPA.

1865. Moore, Proc. Zoöl. Soc. Lond. 780: **Gopala**, Sambara, Bhagava.
Gopala may be taken as the type.

952. SATYRITES* (fossil).

1872. Scudd., Rev. Mag. Zoöl. 66: *Reynesii*. Sole species, and therefore type.
There is a name Satyrites, used for a subfamily group of butterflies by Blanchard and Brullé, in 1840; and therefore, in a memoir on fossil butterflies now in press, I have changed this name to Lethites.

953. SATYRUS.*

1810. Latr., Consid. 440: Teucer, Phidippus, Sophoræ, Piera, *Galathea*, Mæra.
These are all given as types only by Latreille; and it will be seen by comparison of the context that he intended to embrace within it all the Oreades. In a previous work (Sonnini's Buffon), he has placed all of these under his division Satyri of Nymphalis (q. v.); and in the list of names occurs Mæra (given here as one of the types of Satyrus), but it bears there the name of Satyrus (le Satyre of old authors). The name, then, is based upon a synonyme of one of the species included in the group (one of the specified types, indeed), and must therefore be dropped. Moreover, the name is preoccupied in Mammals (Tulp. 1692), and, through Satyra, in Diptera (Meig. 1803).

The subsequent history of the name is as follows: —

1819. God., Encycl. méth. 460: uses it for all the Satyrids.
1822–23. Swains., Zoöl. Ill. i. iii. pl. 159: specifies "Hyperanthus, Galathea, Semele, etc.," as types. If the name could stand, Galathea would then be type. See Agapetes.
1832. Boisduval (loc. var.): most of the European Satyrids.
1851. Westw., Gen. Diurn. Lep. 388: specifies Semele and Fidia as representative.
1858. Ramb., Cat. Lep. Andal. 25: employs it for Arethusa and other species not given by Latreille.
1867. Butl., Entom. iii. 279: says that the "Satyrus of Godart cannot be used, as the type of that genus was Constantia of Cramer, — a species previously used by Hübner as the type of his genus Hipio."

Here are three errors, two of them based on the untenable theory that an author's first species must be taken as his type, which would be an *ex post facto* rule of great undesirability, and having no proper authority.

1868. Ib., Ent. Monthl. Mag. iv. 194; Cat. Sat. 59: specifies Actæa as type.
1872. Crotch, Cist. Ent. i. 91: erroneously refers the name back to Fabricius [Satyri], and says that Latreille (1805) fixed Megæra as the type.

954. SCADA.

1871. Kirb., Syn. Cat. 23: **Phyllodoce**, Leptalina, Reckia, Philemon, Ethica, Theapbia, Xanthina, Zibia.

As this name is proposed to supplant Salacia (q. v.), Phyllodoce must be taken as the type. See Oleria.

955. SCALIDONEURA.

1871. Butl., Proc. Zoöl. Soc. Lond. 250: **Hermina**. Sole species and designated type.

956. SCELOTHRIX.*

1858. Ramb., Cat. Lép. Andal. i. 63: carthami, Alveus, serratulæ, onopordi, Fritillum, malvæ (Alveolus, melotis), Galactites, cynaræ, carlinæ, cirsii, cacaliæ, centaureæ.

The name falls before Hesperia. See also Pyrgus and Syrichtus.

957. SCHŒNIS.

1816. Hübn., Verz. 28: **Cinxia** (Delia, Cinxia). Sole species, and therefore type.

Used in same manner by Stephens (1850) and Kirby (1858). See also Mellicta.

958. Scolitantides.*

1816. Hübn., Verz. 68: Battus, Hylas.
1869. Butl., Cat. Fabr. Lep. 167: the same.
   The name falls before Rusticus. See also Lycæides.

959. Scoptes.*

1816. Hübn., Verz. 111: Alphæus (Alpheus), Protumnus [also given in same work under Thestor!), Crotopus [also given in same work under Eusalasia!].
1866. Butl., Cat. Fabr. Lep. 176: employs it for Alphæus (Alpheus) only.
   This, however, cannot be taken as type, as it had been previously selected as the type of Capys. Protumnus has been chosen as the type of Thestor, and Crotopus belongs to a distinct subfamily. Owing to the somewhat heterogeneous nature of the group, and the fact that two out of the three species were also placed elsewhere by Hübner, the name may as well be dropped. See Capys.

960. Scudderia.*

1873 (Aug.). Grote, Can. Ent. v. 144: *Antiopa*. Sole species and designated type.
   The name falls before Papilio, previously restricted to this species, and is preoccupied in Orthoptera (Stål, April, 1873).

961. Semelia.

1844. [Boisd. in] Doubl., List Br. Mus. 64: **Vibilia**, Aliphera.
1870. Boisd., Lép. Guat. 35: claims the name, mentioning only Vibilia, which therefore becomes the type.
   The name is very close to Semele (Schum., Moll. 1817).

962. Semicaudati.*

1860. Koch, Stett. Ent. Zeit. xxi. 231: Nireus, and a number of other swallow-tails, having no sort of distinctive character but the comparative length of their tails.
   The formation of the name is itself objectionable, and the appearance of such divisions as the semicaudati, caudati, and ecaudati of this author, less than half a generation ago, is an extraordinary case of the "survival" of the spirit of mediæval science. The group is mentioned here only to make this historical sketch complete.

963. Semomesia.

1851. Westw., Gen. Diurn. Lep. 455: **Crœsus**, geminus.
   Crœsus may be taken as the type.

964. SERICINUS.

1851. Westw., Trans. Ent. Soc. Lond. [N. S.] i. 173: **Telamon.** Sole species and designated type.
1852. Ib., Gen. Diurn. Lep. 530: the same.
1856. Gray, Pap. Brit. Mus. 78; Cat. Pap. 93: Telamon and others.

965. SETABIS.

1847. Doubl., List Br. Mus. 19: **Myrtis,** Mæonis [both species inedited].
1851. Westw., Gen. Diurn. Lep. 450: Myrtis, Serica.

Both are described, and the latter figured. It would be better, however, to designate Myrtis as the type, as one of those specified by Doubleday. Mæonis, however, may be the same as Serica, as it appears to be hitherto only a MS. name.

966. SETODOCIS.

1820. Billb., Enum. Ins. 78: Philomela (Lisandra), Dejanira, Mineus, **Peribæa** (Peribœa), Phedra, Hesione (Ocirrhoe).

Peribæa may be taken as the type.

967. SICYONIA.

1816. Hübn., Verz. 13: **Rhea** (Sara, Thamar), Apseudes, Erato (Doris).

Rhea may be taken as the type. See Laparus.

968. SIDERONE.

1822–26. Hübn., Exot. Schmett. ii.: **Ide.** Sole species, and therefore type.
1836. Boisd., Spec. gén., pl. 4 B.: the same.

Subsequently used in the same sense by Doubleday, Westwood, Felder, Kirby, etc. See also Phyllophasis.

1870. Boisd., Lép. Guat. 51: employs it for Mars and Isidora, previously placed in the same group by other authors, and quotes the genus as his own! yet, in 1836, he uses it for the sole species placed in it by Hübner!

969. SIPROETA.

1822–26. Hübn., Exot. Schmett. ii.: **Trayja.** Sole species, and therefore type. See Amphirene.

970. SIRONIA.

1823. Hübn., Zutr. ii. 31: **Tithia.** Sole species, and therefore type.

971. SISEME.

1851. Westw., Gen. Diurn. Lep. 462: Aristoteles, **Electryo**.
1867. Bates, Journ. Linn. Soc. Lond. ix. 433: the same and others.
1871. Kirb., Syn. Cat. 309: the same.

Electryo, having been figured by Westwood, may be taken as the type.

972. SITION.

1816. Hübn., Verz. 77: **Nedymond**, Melampus.
1866. Trim., Rhop. Afr. Austr. 232: employs it for Anta (Batikeli), allied to Melampus.
1871. Kirb., Syn. Cat. 411: employs it for Nedymond and its allies, excluding Melampus.

Melampus was taken in 1863 by Hewitson to form his Deudorix, and therefore we may follow Kirby in considering Nedymond as the type.

973. SMYRNA.

1822-26. Hübn., Exot. Schmett. ii.: **Blomfildia** (Blomfildii). Sole species, and therefore type.
1827-37. Gey. in Hübn., Exot. Schmett. iii.: uses it for Karwinskii.

Westwood, Felder, and Kirby use it for both these species.

974. SOSPITA.

1861. Hewits., Exot. Butt. ii. 91: Tantalus, Savitri (Susa), Neophron, Segecia, **Fylla**, Echerius, Tepahi.
1861. Herr.-Schaeff., Exot. Schmett. pt. 37: employs it for the first four of the above.

Fylla, being generically distinct from Echerius the type of Abisara, may be selected to represent this genus.

975. SPATHILEPIA.

1870. Butl., Ent. Monthl. Mag. vii. 57: Tamyroides, **Clonius**, Cellus. Clonius specified as type.

Used by Kirby in the same sense.

976. SPEYERIA.

1872. Scudd., Syst. Rev. 23: **Idalia**. Sole species and designated type.

977. SPHÆNOGONA.

1870. Butl., Cist. Ent. ii. 35, 44: **Ectriva**, bogotana. Ectriva is specified as type; it was undescribed until later, but before further use of the generic name.

978. Spilothyrus.*

1832. Dup., Pap. France, Diurn. Suppl. 415: alceæ (malvæ), altheæ (althea), lavateræ.
1858. Ramb., Cat. Lép. Andal. 79: employs it for the same.
1861. Staud., Cat. 14: the same.

The name falls before Urbanus. See also Carcharodus.

979. Spindasis.

1857. Wallengr., Rhop. Caffr. 45: **natalensis** (Masilikazi). Sole species, and therefore type.

980. Spioniades.

1816. Hübn., Verz. 114: **Artemides**, Alcmon (Almon), Psecas.

Artemides may be taken as the type.

981. Stalachtis.

1816. Hübn., Verz. 27: **Phlegia**, Euterpe, Phædusa, Calliope.
1847. Doubl., List Br. Mus. 19: the same and others.
1848. Ib., Gen. Diurn. Lep. 133: uses it for Calliope, Euterpe, Susanna, and Phlegia.
1851. Westw., Gen. Diurn. Lep. 466: employs it for eight species, including all of Hübner's.
1867. Bates, Journ. Linn. Soc. Lond. ix. 457: extends it still further, employing also the Hübnerian species.
1871. Kirb., Syn. Cat. 333: uses it in the same sense.

Phlegia may be selected as the type of this genus, which is distinct from Nerias.

982. Steroma.

1851. Westw., Gen. Diurn. Lep. 400: **Bega**. Sole species, and therefore type, as stated by Butler.

983. Steropes.*

1832. Boisd., Voy. Astrol. 167: picta, ornata, Iacchus (Jacchus).
1836. Ib., Spec. gén., pl. 9 B.: uses it for Palæmon (Paniscus) only.

As the name is derived from that of one of the species intended, and afterwards employed by Boisduval as one of this group, it must fall. It is also preoccupied in Coleoptera (Stev. 1806). See Carterocephalus and Pamphila.

984. Sterosis.*

1865. Boisd. in Feld., Reise Novara, 219: *Brassolis* (robusta). Sole species, and therefore type.

The name falls before Liphyra (q. v.).

985. STIBOCHIONA.

1868. Butl., Proc. Zoöl. Soc. Lond. 614: Nicea, **Coresia**. Coresia specified as type.

986. STICHOPHTHALMA.

1862. Feld., Wien. Ent. Monatschr. vi. 27: **Howqua**. Sole species, and therefore type.

987. STOMYLES.

1872. Scudd., Syst. Rev. 55: **textor**. Sole species and designated type.

988. STRYMON.

1816. Hübn., Verz. 74: **Titus** (Mopsus), pruni, betulæ, w. album, ilicis (esculi, ilicis), acaciæ, Melinus, spini (Lynceus, spini), Beon, Pan, Acis (Mars).
1850. Steph., Cat. Brit. Lep. 16, 260: places here betulæ, pruni, w. album, Titus, spini, and ilicis.
1858. Kirb., List Brit. Rhop.: uses it for only pruni, w. album, spini, and ilicis; but Thecla was restricted much earlier.
1869. Butl., Cat. Fabr. Lep. 190: employs it (sensu stricto) for eight species, including, of Hübner's, Titus, pruni, w. album, ilicis, and spini.
1872. Scudd., Syst. Rev. 32: specifies Titus as type, which follows from the action of Stephens and Butler.

989. STYGNUS.*

1867. Feld., Reise Novara, 489: *humilis*. Sole species, and therefore type, as specified by Butler.

But the name is preoccupied in Arachnids (Perty, 1830).

990. SUNIAS.

1816. Hübn., Verz. 12: Phyllis, **Melpomene** (Lucia, Melpomene, Callicopis).

Melpomene may be taken as the type. See also Phlogris and Laparus.

991. SYMBRENTHIA.

1816. Hübn., Verz. 43: **Hyppoclus** (Hippocle). Sole species, and therefore type.
1871. Kirb., Syn. Cat. 180: the same and others.

See Laogona.

992. SYMETHA.*

1828. Horsf., Descr. Cat. Lep. E. Ind. Co. 59, expl. pl. 2: *Symethus* (Pandu). Sole species, and therefore type.
1832. Boisd., Voy. Astrol. 72 [Simœthus]: Rex, Pandu.

The name, being derived from that of the species upon which it is grounded, falls. It is also preoccupied in Crustacea (Rafin. 1814). See Gerydus and Miletus.

993. SYMMACHIA.

1816. Hübn., Verz. 26: Helius (Ochima), **Probetor** (Probetrix).
1837. Sodoffsk., Bull. Mosc. x. 82: not knowing that this name was already in use, proposes to employ it in place of Hesperia.
1847. Doubl., List Br. Mus. 8: employs it for Probetor and others.
1851. Westw., Gen. Diurn. Lep. 444: makes a similar but more extended use of it.
1867. Bates, Journ. Linn. Soc. Lond. ix. 437: extends it still further in the same sense.
1871. Kirb., Syn. Cat. 313: the same.

Probetor is therefore the type.

994. SYMMACHLAS.

1820-21. Hübn., Exot. Schmett. ii.: **nigrina**. Sole species, and therefore type.
1821. Ib., Index, 5: nigrina.

995. SYMPHÆDRA.

1816. Hübn., Verz. 40: Æropus (Ærope), **Thyelia** (Alcandra), Evelina, Lysandra.
1844. Doubl., List Br. Mus. 105: [Symphedra]. Employs it for Thyelia and unnamed species only.
1850. Westw., Gen. Diurn. Lep. 294: Thyelia, Æropus.

Used in same sense by subsequent authors. Thyelia becomes the type, through Doubleday's action.

996. SYNALPE.*

1870. Boisd., Lép. Guat. 36: *Thirza* (Euryale). Sole species, and therefore type.

Falls before Anelia. See also Clothilda.

997. SYNAPTA.*

1865. Feld., Reise Novara, 294: *Arion*. Sole species, and therefore type.

The name is preoccupied in Echinoderms (Eschsch. 1829).

998. SYNARGIS.

1816. Hübn., Verz. 18: Phyleus (Phyllea), Orestes (Orestessa), Soranus (Sorane), **Tytia**, Odites (Oditis).

Tytia may be taken as the type.

999. SYNCHLOE.

1816. Hübn., Verz. 94: Callidice, Autodice, Hellica, Chloridice, Daplidice, **Belemia** (Belemia, Glauce).
1844. [Boisd. in] Doubl., List Br. Mus. 76: employs this name for Erodyle, Janais, Narva (Bonplandi), and some MS. species, all of which have nothing whatever to do with Hübner's group. See Chlosyne.
1848. Boisd. in ib., Gen. Diurn. Lep. 185: follows the same course.
1858. Kirb., List Brit. Rhop.: first restores the Hübnerian sense by employing it for Daplidice, which would therefore become the type, but that it had already been taken as the type of Pontia.
1861. Feld., Neues Lep. 10: follows Doubleday.
1870. Butl., Cist. Ent. i. 38, 51: specifies Callidice as type, but wrongly.
1872. Scudd., Syst. Rev. 42: does the same, with equal error.

All the species but Belemia having been taken either for Pontia or Tatocheila (q. v.), this becomes the type.

1000. SYNGEA.*

1816. Hübn., Verz. 62: Arachne (Pronoe, Pitho), Alecto.

The name falls before Erebia. See also Gorgo, Marica, Phorcis, Epigea, and Oreina.

1001. SYNPALAMIDES.*

1822–26. Hübn., Exot. Schmett. ii.: *Mimon*. Sole species, and therefore type.

It is not a butterfly.

1002. SYRICHTUS.*

1832–33. Boisd., Icones, 230: Proto, Sao (Therapne), Orbifera (Orbifer), cacaliæ (alveus), Fritillum, Tessellum, malvæ (malvæ, Alveolus), alceæ (alceæ, malvarum), lavateræ, sidæ, carthami, altheæ.

The name of the group is derived from that of one of the species which Boisduval must have intended to include in it, and therefore falls. The first four only are described in the Icones: the others are only alluded to in his remarks on the genus. See Hesperia, Pyrgus, and Scelothrix.

1003. SYRMATIA.

1816. Hübn., Verz. 23: **Dorilas** (Nyx), Rhetus (Rhete), Aulestes (Aulestis), Chorineus (Chorinea).
1847. Doubl., List Br. Mus. 4: employs the name for Dorilas only, which therefore becomes the type.
1851. Westw., Gen. Diurn. Lep. 426: the same.

Bates and Kirby use it similarly. See Dorila.

1004. TACHYRIS.

1867. Wall., Trans. Ent. Soc. Lond. [3] iv. 361: I. Hombronii, Cardena, Nerissa, Lyncida (Lynceola, Lyncida, formosana, Andrea, Hippo), Enarete, Scyllara (Scyllaria), Ada (Ada, Clavis), abnormis, Panda (Panda, Nathalis), Paulina, Albina (Rouxii), Psyche, Galathea, Ega, Urania, Agave (Jacquinotii), Alope, Amarella, Acrisa, Leptis; II. Celestina, Clementina, Athama, Cynisca, Eumelis, Panthea, Cycinna, Corinna, Liberia, Eliada, Placidia, Fatime (Fatima); III. **Nero** (Nero, Domitia), Galba, Zarinda, bournensis, Zamboanga, Asterope, Ithome, Nephele; IV. Pandione, Lucasii, Indra, Phœbe, Nephele (Zamora), Lalage (Lalago); V. Polisina, Ægis (Illana).
1871. Kirb., Syn. Cat. 463: uses it in the same sense.

Nero may be taken as the type. See Trigonia.

1005. TÆNARIS.

1816. Hübn., Verz. 53: **Urania** (Jaira, Nysa). Sole species, and therefore type.
1865. Herr.-Schaeff., Prodr. 71: employs it for the same and many others.
1871. Kirb., Syn. Cat. 117 [Tenaris*]: follows Herrich-Schaeffer.

See also Drusilla.

1006. TAGIADES.

1816. Hübn., Verz. 108: **Japetus**, Paulinus.
1869. Butl., Cat. Fabr. Lep. 283: employs it for four species, including none of Hübner's.

---

* The word is given in four different ways in Hübner's Verzeichniss: Tænares and its German equivalent Tänaren at the head of the group, where the names are always given in the plural form; Tenaris, in connection with the species; and Tænaris, in the index. This, as well as the derivation of the word, shows that Tenaris was simply a misprint.

1870. Ib., Ent. Monthl. Mag. vii. 99: specifies Japetus as type.
1871. Kirb., Syn. Cat. 634: employs it for both Hübner's specie and many others.

1007. TALIDES.

1816. Hübn., Verz. 106: Athenion, Corbulo (obscurus), **Sergestus** (Sinois), Broteas, Astylos, Celænus, Ramusis.
1869. Butl., Cat. Fabr. Lep. 266: employs it for Nicias, Phocus, Sergestus (Sinon), and Sebaldus, the third of them only one of Hübner's, and therefore type.
1870. Ib., Ent. Monthl. Mag. vii. 93: specifies Athenion as type, but wrongly; his own previous action having determined that Sergestus must be the type, the two species not being strictly congeneric.

1008. TAMYRIS.

1820-21. Swains., Zoöl. Ill. i. i. 33: **Zeleucus**. Sole species, and therefore type.

Subsequently, in the same work, he added other species. See Pachyrhopala.

1009. TANAECIA.

1868. Butl., Proc. Zoöl. Soc. Lond. 610: Calliphorus, Valmikis, Apsarasa, Varuna (supercilia), Varuna, Aruna, Lutala, Trigerta, Pelea, **Pulasara** (Pulasara, Vikrama), Violaria. Pulasara is specified as type.
1871. Kirb., Syn. Cat. 257: uses it similarly.

1010. TANAOPTERA.*

1820. Billb., Enum. Ins. 79: Amalthea (Amathea), Europa, Leda (Leda, Banksii).

This name may be allowed to drop, from the heterogeneous nature of the contents of the genus.

1011. TAPINA.*

1820. Billb., Enum. Ins. 81: proposes this name, for no reason, to supplant Emesis. He gives no species.

1012. TARACTROCERA.

1869. Butl., Cat. Fabr. Lep. 279: **Mævius**. Sole species, and therefore type, as specified later by Butler, and as used by Kirby.

1013. TATOCHEILA.

1870. Butl., Cist. Ent. i. 38, 51: **Autodice** (Autodyce). Sole species and designated type.

Is this genus distinct from Pontia? See also Synchloe.

1014. TAXILA.

1847. Doubl., List Br. Mus. 2: **Haquinus** (Drupadi), Orphna, Echerius, and some MS. species.
1851. Westw., Gen. Diurn. Lep. 421: employs it for all the above and others.
1861. Hewits., Exot. Butt. ii. 91: uses it for the first two of Doubleday's species and a number of others.
1867. Bates, Jour. Linn. Soc. Lond. ix. 414: employs it for Orphna, Haquinus (Drupadi), and others.
1871. Kirb., Syn. Cat. 285: follows Bates.

Haquinus may be taken as the type.

1015. TAYGETIS.

1816. Hübn., Verz. 55: Virgilia, Andromeda (Thamyra, Andromeda), **Mermeria**, Celia.
1851. Westw., Gen. Diurn. Lep. 355: Mermeria is mentioned as "a good type of the genus," and all of Hübner's other species are included in it, besides others.
1865. Herr.-Schaeff., Prodr. i. 58: the same.
1867. Butl., Ent. Monthl. Mag. iv. 194: specifies Virgilia as type.
1871. Kirb., Syn. Cat. 108: uses it for all of Hübner's species and others.

On account of Westwood's statement, Mermeria should be considered the type.

1016. TEINOPALPUS.*

1843. Hope, Trans. Linn. Soc. Lond. xix. 131: *imperialis*. Sole species, and therefore type.

Since used for same species by Doubleday, Gray, and Kirby, but properly objected to by Felder as of mongrel origin. See Teinoprosopus.

1017. TEINOPROSOPUS.

1864. Feld., Spec. Lep. 1: **imperialis**. Sole species, and therefore type.
1867. Herr.-Schaeff., Prodr. ii. 19: the same.

Proposed by Felder to replace Teinopalpus (q. v.). "Pristinum nomen vox hybrida."

1018. TELCHIN.*

1825. Hübn., Cat. Franck, 85: *Licus* and three MS. species. Licus is therefore type.

It is given with the authority Cramer, and is doubtless meant for Castnia Lycas of Verlorens's Catalogue of Cramer.

As it is not a butterfly, we have nothing to do with it in this place.

1019. TELCHINIA.

1816. Hübn., Verz. 27: violæ (Cephea), Medea (Saronis), Cæcilia (Bendis), Zetes (Mycenæa, Zetis), Horta, **Serena**, vesta (Issoria).

1848. Doubl., Gen. Diurn. Lep. 141: employs it for a number of species, including, of Hübner's, violæ, Cæcilia, and Serena.

1857. Horsf.-Moore, Cat. Lep. E. Ind. Co. i. 135: adopt Doubleday's restriction, and employ it for violæ only.

This would therefore become the type, were it not probable that it is strictly congeneric with Horta, the type of Acræa. Serena may be chosen.

1020. TELEGONUS.

1816. Hübn., Verz. 104: Talus, Phocus (Phocus, Morpheus), **Anaphus**, Midas.

1869. Butl., Cat. Fabr. Lep. 261: employs it for all of the above, excepting Phocus, and for others.

1870. Ib., Ent. Monthl. Mag. vii. 56: specifies Talus as the type.

1871. Kirb., Syn. Cat. 572: uses it in this sense.

Talus, however, cannot be taken as the type, as it belongs to Thymele, whose type was earlier established. Anaphus may be taken as the type.

1021. TELEMIADES.

1816. Hübn., Verz. 106: **Avitus**, Epicalus, Salatis.

1869. Herr.-Schaeff., Prodr. iii. 68: employs it for Avitus and others. Avitus therefore becomes the type.

1022. TELESTO.*

1832. Boisd., Voy. Astrol. 164: *Peronii* (Perronii). Sole species, and therefore type.

1862. Feld., Verh. Zoöl.-bot. Gesellsch. Wien. xii. 491: describes three new species, recognizing Peronii as the type.

1869. Herr.-Schaeff., Prodr. iii. 53: without indication of species.

1870. Butl., Ent. Monthl. Mag. vii. 96: specifies Dirpha as type, of course erroneously.

The name is preoccupied in Polyps (Lamx.-1812) and Crustacea (Rafin. 1814). See Hesperilla.

1023. TEMENIS.

1816. Hübn., Verz. 34: Minerva (Arcadia), **Laothoe** (Merione), Erigone, Hedonia, Laomedia.
1871. Kirb., Syn. Cat. 204: employs it for Sylphis, pulchra, and Laothoe.
   Laothoe should therefore be considered as the type.

1024. TERACOLUS.

1832–33. Swains., Zoöl. Ill. ii. 115: **subfasciatus**. Sole species, and therefore type, as stated by Butler.
   Used in same sense by Kirby. See also Ptychopteryx and Thespia.

1025. TERIAS.

1820–21. Swains., Zoöl. Ill. i. 22: Elvina, **Hecabe**. Hecabe designated as type.
1836. Boisd., Spec. gén. 651: employs it for the two above-mentioned and more than fifty others.
   It is similarly used by Doubleday and others.
1870. Butl., Cist. Ent. i. 35, 44: specifies Hecabe as the type.

1026. TERINOS.*

1836. Boisd., Spec. gén., pl. 5 B.: *Clarissa*. Sole species, and therefore type.
   Used in same sense by Doubleday, Felder, and Kirby, but the name is preoccupied through Terina (Hübn., Lep. 1816).

1027. TETRAGONUS.*

1832. Gey. in Hübn., Zutr. iv. 17: *Catamitus*. Sole species, and therefore type.
   According to Westwood (Gen. Diurn. Lep. 504), it is not a butterfly. The name is in any case preoccupied through Tetragonum (Quoy et Gaim. 1824) and Tetragona (Ib. 1827).

1028. TETRAPHLEBIA.

1867. Feld., Reise Novara, 487: **Germainii**. Sole species, and therefore type, as stated by Butler.

1029. THAIS.

1807. Fabr., Ill. Mag. vi. 283: Polyxena (Hypsipyle). Sole species, and therefore type.
1810. Latr., Consid. 440: specifies Rumina as type, but wrongly.

1815. Oken, Lehrb. i. 726: employs it for Polyxena.
1816. Hübn., Verz. 89: employs it for both of the above.
 It is subsequently used in the same sense by all authors, but is preoccupied in Mollusca (Bolt. 1798). See Zerynthia and Eugraphis.

1030. THALEROPIS.

1871. Staud., Cat. Eur. Lep. 17. **Ionia.** Sole species, and therefore type.
1871. Kirb., Syn. Cat. App. 649: the same.

1031. THANAOS.

1832–33 (either late in 1832 or early in 1833). Boisd., Icon. 240: Marloyi, **Tages.**
1833–34 (probably 1834 or late in 1833). Boisd.-LeC., Lép. Amér. Sept., pl. 65, 66: Juvenalis, Brizo.
1836. Boisd., Spec. gén., pl. 9 B.: Tages.
1870. Butl., Ent. Monthl. Mag. vii. 97: specifies Juvenalis as type, but wrongly.
 Tages should be taken as the type. See Nisoniades and Erynnis.

1032. THAROPS.

1816. Hübn., Verz. 109: **Menander,** Thersander. [Placed by Hübner among the Urbicolæ!]
1847. Doubl., List Br. Mus. 14: employs it for Menander only, which thereby becomes the type.
 Used in this same sense by subsequent authors.

1033. THAUMANTIS.

1822–26. Hübn., Exot. Schmett. ii.: **Odana** (Oda). Sole species, and therefore type.
1836. Boisd., Spec. gén., pl. 8 B.: the same.
 Subsequently used by Doubleday, Westwood, and Kirby in the same sense.

1034. THECLA.

1807. Fabr., Ill. Mag. vi. 286: betulæ, **spini,** quercus.
1815. Oken, Lehrb. i. 721: employs it for the same and others.
1815. Leach, Edinb. Encycl. 718: uses it for betulæ, pruni (generically identical with spini), and quercus.
1821–22. Swains., Zoöl. Ill. i. ii. 69: specifies betulæ as type.
 Subsequent authors have employed it for the whole body of European hair-streaks.

1829. Curtis, Brit. Ent. pl. 264: designates betulæ as the type.
1840. Westw., Gen. Syn. 88: does the same.
1872. Crotch, Cist. Ent. i. 66: says that betulæ is type, through Dalman in 1816; but Dalman specifies betulæ as type of Zephyrus, of which Aurotis was a section.
1870. Kirb., Journ. Linn. Soc. Lond. Zoöl. x. 499: says "it would be far more convenient and quite justifiable" to take spini as type.
1872. Scudd., Syst. Rev. 29: specifies spini as type.

Betulae cannot be taken as the type on account of the foundation in 1816 of Dalman's Zephyrus, and consequently spini must be chosen.

1035. THEMONE.

1851. Westw., Gen. Diurn. Lep. 461: **Pais**. Sole species in the typical section and designated type of the genus.
1867. Bates, Journ. Linn. Soc. Lond. ix. 425: employs it for Pais and two others.
1871. Kirb., Syn. Cat. 299: uses it in a similar manner.

1036. THEOPE.

1847. Doubl., List Br. Mus. 6: Lagus, **Terambus**, and some MS. species.
1851. Westw., Gen. Diurn. Lep. 439: employs it for several species, including only Terambus (Lytæa, Terambus) of Doubleday's, which thereby becomes the type.
1858. Moore, Cat. Lep. E. Ind. Co. i. 234: uses it (as new) for Himachala, an entirely different insect. See Anadebis.
1868. Bates, Journ. Linn. Soc. Lond. ix. 453: uses it for Terambus and very many others.
1871. Kirb., Syn. Cat. 330: uses it for many species, including Terambus.

See also Psalidopteris.

1037. THEOREMA.

1865. Hewits., Ill. Diurn. Lep. 69: **Eumenia**. Sole species, and therefore type.

Used for same species by Kirby.

1038. THEREUS.

1816. Hübn., Verz. 79: **Lausus**. Sole species, and therefore type.

1039. THERITAS.

1816. Hubn., Verz. 80: imperialis (Venus), **Mavors**.

1869. Butl., Cat. Fabr. Lep. 194: employs it for Actæon, with which he places imperialis (Venus).

>This, therefore, would become the type, but imperialis became in 1832 the type of Arcas, leaving Mavors for the type of this group.

1040. THERIUS.*

1820. Dalm. in Billb., Enum. Ins. 75: Apollo, Mnesonyme.

>The name is preoccupied through Theria (Hübn., Lep. 1816) and Thereus (Ib.). See Parnassius and Doritis.

1041. THESPIA.*

1858. Wallengr., K. Vet. Akad. Förh. xv. 77: *Bohemanni.* Sole species, and therefore type.

>Doubtless intended by Wallengren to supplant his Ptychopteryx (preoccupied), but it falls before Teracolus.

1042. THESTIAS.*

1836. Boisd., Spec. gén. 590: Pyrene (Ænippe, Pirene), Marianne, Vollenhovii (Balice), Venilia.

>· Subsequently used by Doubleday and others, but the name is preoccupied through Thestius (Hübn., Lep. 1816). See Ixias.

1043. THESTIUS.

1816. Hübn., Verz. 78: Gabriela (Gabrielis), **Pholeus** (Pholeus, Archytes), Hyacinthus, Ematheon, Lycabas (Lycabus).

>Pholeus may be taken as the type.

1044. THESTOR.

1816. Hübn., Verz. 73: **Protumnus** (Petalus), Ballus.
1857. Led., Wien. Ent. Monatschr. i. 32: employs it for Ballus and others, not including Protumnus.
1861. Staud., Cat. Lep. Eur. 3: follows Lederer.
1869. Butl., Cat. Fabr. Lep. 174: uses it for Protumnus and others.
1871. Kirb., Syn. Cat. 345: follows Lederer.

>Ballus, however, cannot be taken as the type, since previously to Lederer's action it had been selected as the type of Tomares (q. v.); we must therefore follow Butler in considering Protumnus as the type.

1045. THISBE.

1816. Hübn., Verz. 24: **Irenæa** (Belise). Sole species, and therefore type.

>Since used similarly by Bates and Kirby.

1046. THOAS.*

1832–33. Swains., Zoöl. Ill. 121: Hectorides (Lysithous), Thoas, Agavus (Agavius), Pompeius (Paris), Androgeos (Androgeus). Thoas and Agavus are specified as typical.

Since the name is founded on one of the typical species, it must drop. See Heraclides.

1047. THORYBES.

1872. Scudd., Syst. Rev. 50: **Bathyllus**, Pylades, Nevada. Bathyllus specified as type.

1048. THRACIDES.

1816. Hübn., Verz. 105: **Phidon**, Salius.
1869. Herr.-Schaeff., Prodr. iii. 44: employs the name, but without specification. Kirby (Syn. Cat. 624) credits him with placing here some of the species of Butleria Kirby, none of which are older than 1852.
1871. Kirb., Syn. Cat. 578: uses it for both of Hübner's species and for others.

Phidon may be taken as the type.

1049. THRENODES.*

1870. Hewits., Equat. Lep. iv. 58: *Cænoides*. Sole species, and therefore type.
1871. Kirb., Syn. Cat. 306: the same.

The name is preoccupied in Lepidoptera (Duponch. 1844.). See Nahida.

1050. THYCA.

1858. Wallengr., K. Vet. Akad. Förh. xv. 76: I. Hyparete, Egialea; II. **Aganippe**.
1869. Butl., Cat. Fabr. Lep. 205: employs it for the species in Wallengren's first section, and for others.

But these must be placed in Delias, and consequently Aganippe must be taken as the type.

1051. THYMELE.

1807. Fabr., Ill. Mag. vi. 287: I. Proteus, **Mercatus**, Apastus (Acastus); II. Thrax, Gnetus, Bixæ; III. Morpheus (Aracinthus), malvæ, Tages.
1815. Oken, Lehrb. i. 758: employs it for Proteus, Mercatus, Apastus (Acastus), and others not mentioned by Fabricius.

1828. Steph., Ill. Brit. Ent. Haust. i. 97: restricts it to malvæ, Tages, and others not mentioned by Fabricius.
1840. Westw., Gen. Syn. 88: specifies Tages as type.
1871. Kirb., Syn. Cat. 569: Proteus and its immediate allies.
1872. Scudd., Syst. Rev. 47: specifies Proteus as type.

Proteus, however, cannot be taken as type; for in 1832 it became the type of Eudamus. By Oken's action the genus must be restricted to Fabricius's first section after the removal of Proteus, and Mercatus may be taken as the type.

1052. THYMELICUS.

1816. Hübn., Verz. 113: Actæon, Pustula, **Vibex**, Thaumas (Venula, linea), lineola (Virgula), Vitellius, Numitor (Puer).
1850. Steph., Cat. Brit. Lep. 22: employs it for Actæon and Thaumas (linea).
1858. Kirb., List Brit. Rhop. [Thymelinus]: uses it for Actæon only.
1869. Herr.-Schaeff., Prodr. iii. 44: uses it without specification of members.
1870. Butl., Ent. Monthl. Mag. vii. 94: specifies Actæon as type.
1871. Kirb., Syn. Cat. 609: uses it in this sense.
1872. Scudd., Syst. Rev. 54: specifies Actæon as type.

Thaumas, however, is the type of Adopæa, and Actæon and lineola belong to the same genus. Vitellius belongs to Atrytone (1872), and Numitor is the type of Ancyloxypha (1862). Pustula and Vibex remain: these belong to Hedone (1872), which may fall before this name. Vibex may be taken as the type.

1053. THYRIDIA.

1816. Hübn., Verz. 9: Themisto, **Psidii**, Ilione.
1844. Doubl., List Br. Mus. 59: employs it for the same and others.
1847. Ib., Gen. Diurn. Lep. 117: uses it for Psidii and Ædesia.
1862. Bates, Linn. Trans. xxiii. 519: employs it for Pytho (Ino) only.
1864. Herr.-Schaeff., Prodr. i. 47: employs it for several, including only Psidii of Hübner's list.
1870. Boisd., Lép. Guat. 80: employs it for Eupompe, etc. See his Xanthocleis for this group.
1871. Kirb., Syn. Cat. 19: uses it for Themisto, Psidii, and others not of Hübner's list.

Psidii becomes the type by Doubleday's action in 1847.

1054. THYSONOTIS.

1816. Hübn., Verz. 20: **Danis**, Athemon (Athemæna).
1860. Feld., Wien. Ent. Monatschr. iv. 224: employs it (as new?) for Inops and others related to Danis.

> Athemon is the type of Eubagis, and Danis may be taken as the type. See Damis and Danis.

1055. TIGRIDIA.

1816. Hübn., Verz. 40: **Aceste**, Dirce, Zingha.
1844. Doubl., List Br. Mus. 93: employs it for Aceste only, which therefore becomes the type.

> See Callizona.

1056. TIMETES.

? 1836. Boisd. in Cuv., Règne An. Ed. Disc. ii., pl. 139 [Tymetes]: **Merops**. Sole species, and therefore type.
1844. Doubl., List Br. Mus. i. 87: Coresia, Themistocles, Chiron, Orsilochus, Corinna, and some unpublished species.
1850. Westw., Gen. Diurn. Lep. 262: employs it for all the above and for others.
1870. Boisd., Lép. Guat. 44: uses it for Corinna and others.

1057. TINGRA.

1847. Boisd., Voy. Deleg. ii. 589: **tropicalis**. Sole species, and therefore type.
1852. Westw., Gen. Diurn. Lep. 504: refers to it as probably allied to Pentila.
1857. Wallengr, Rhop. Caffr. 46: the same.

> See Pentila and Liptena.

1058. TISIPHONE.

1816. Hübn., Verz. 60: **Abeona** (Zelinde), Pasiphae (Pasyphae), Tulbaghia (Tulbachii).
1822–26. Ib., Exot. Schmett. ii.: Hercyna.
1844. Doubl., List Br. Mus. 150: Hercyna.
1851. Westw., Gen. Diurn. Lep. 370: the same.
1865. Herr.-Schaeff., Prodr. i. 61: Hercyna and another.
1868. Butl., Ent. Monthl. Mag. iv. 194: specifies Hercyna as type.
1868. Ib., Cat. Sat. 71: the same; but refers the genus to Westwood, and adds in a note that Abeona is the type of Hübner's Tisiphone.

1871. Kirb., Syn. Cat. 46: Hercyna. He queries which of Hübner's references is the older, but there can be little doubt upon the point, thanks to Hübner's Index.

Hercyna cannot be taken as the type, as it is not congeneric with any of the species upon which the genus was founded, nor is it one of the original list. Tulbaghia became type of Meneris in 1844. Pasiphae belongs to Pyronia (1816), so that Abeona must be taken as the type. See also Heteronympha and Hipparchioides.

1059. TITHOREA.

1847 (June). Doubl., Gen. Diurn. Lep., pl. 14: Bonplandi, **Harmonia** (Megara).
1847 (Aug.). Ib., ib. 99: I. Humboldtii, Bonplandi, Pavonii; II. Irene, Harmonia (Megara), Tyro.
1862. Bates, Linn. Trans. xxiii. 552: employs it for Harmonia and a new species.
1864. Herr.-Schaeff., Prodr. i. 50: uses it much as Doubleday did.
1871. Kirb., Syn. Cat. 35: the same.

Harmonia is the type through Bates.

1060. TMETOGLENE.

1862. Feld., Wien. Ent. Monatschr. vi. 235. **Esthema**. Sole species, and therefore type.

Used in same way by Bates, Herrich-Schaeffer, and Kirby. See Brachyglenis.

1061. TMOLUS.

1816. Hübn., Verz. 76: Megacles, Sylvanus (Syllidus), Crolus, **Echion**, Eurytulus.
1869. Butl., Cat. Fabr. Lep. 187: employs it for Echion and several others not specified by Hübner.

Echion therefore becomes the type.

1062. TOMARES.

1839. Ramb., Faune Ent. Andal. ii. 261: **Ballus**. Sole species, and therefore type.
1871. Kirb., Syn. Cat. 345 [Thomares]: given as a synonyme of Thestor (q. v.).

1063. TRAPEZITES.

1816. Hübn., Verz. 112: **Symmomus**. Sole species, and therefore type, as stated by Butler.
1869. Herr.-Schaeff., Prodr. iii. 49: used without mention of species.
1871. Kirb., Syn. Cat. 621: used in same sense.

1064. TREPSICHROIS.

1816. Hübn., Verz. 16: **Midamus** (Basilissa, Mulcibra, Midamis), Alca, Eleusina.
  Midamus may be taken as the type.

1065. TRICHONIS.

1865. Hewits., Ill. Diurn. Lep. 68: **Theanus**. Sole species, and therefore type.
1871. Kirb., Syn. Cat. 427: the same.

1066. TRIGONIA.*

1837. Gey. in Hübn., Zutr. v. 21: *Nero*. Sole species, and therefore type.
  The name is preoccupied in Mollusks (Brug. 1791). See Tachyris.

1067. TRIOPADES.

1816. Hübn., Verz. 73: Orus, **Eupalemon**.
  Eupalemon may be taken as the type. This species is wrongly placed by Kirby among the Urbicolæ.

1068. TRIPHYSA.

1850. Zell., Stett. Ent. Zeit. 308: Dohrnii, **Phryne** (Tircis).
1861. Staud., Cat. 14: employs it for Phryne and Sunbecca.
1865. Herr.-Schaeff., Prodr. i. 60: the same.
1867. Butl., Ent. Monthl. Mag. iv. 194: designates Phryne as the type.
  Is this name too close to Triphassa (Hübn., Lep. 1816) to be used? See Phryne.

1069. TRITONIA.*

1832. Gey. in Hübn., Zutr. iv. 25: *Eupompe*. Sole species, and therefore type.
  This name is preoccupied in Mollusks (Cuv. 1798).

1070. TROIDES.

1816. Hübn., Verz. 88: Priamus, **Helena** (Amphimedon, Helena), Pompeus (Astenous, Minos), Amphrysus, Hippolytus (Remus).
  Helena may be taken as the type. See Amphrisius.

1071. TROILIDES.

1822–26. Hübn., Exot. Schmett. ii: **Torquatus** (Tros). Sole species, and therefore type.

1072. Tyanitis.*

1847. Doubl., List Br. Mus. 19: *Tenes.* Sole species, but undescribed.

The genus also being undescribed, the name falls.

1073. Udranomia.*

1870. Butl., Ent. Monthl. Mag. vii. 58: *Orcinus.* Sole species, and designated type.
1871. Kirb., Syn. Cat. 579: the same.

See Hydrænomia, which supplants it on orthographic grounds.

1074. Uraneis.

1867. Bates, Journ. Linn. Soc. Lond. ix. 411: **hyalina.** Sole species, and therefore type.
1871. Kirb., Syn. Cat. 333: the same.

Is this name too close to Urania (Fabr., Lep. 1807).

1075. Urbanus.

1806. Hübn., Tent. 1: **alceæ** (malvæ).* Sole species, and therefore type.

See Carcharodus, Erynnis, and Spilothyrus.

1076. Utica.

1865. Hewits., Ill. Diurn. Lep. 56: **Onycha.** Sole species, and therefore type.

Thus used by Kirby. Preoccupied in Crustacea (White-Ad. 1847).

1077. Valeria.*

1829. Horsf., Descr. Cat. Lep. E. Ind. Co. 139: *Valeria.* Sole species, and therefore type.

The name, being founded on that of the sole species, falls.

1078. Vanessa.

1807. Fabr., Ill. Mag. vi. 281: Io, **Atalanta,** urticæ, Levana.
1810. Latr., Consid. 440: specifies Atalanta as type.
1815. Oken, Lehrb. i. 729: employs it for Arsinoe and others.
1816. Hübn., Verz. 33: uses it for Huntera (Hunteri), Carye, and cardui.
1825. Curtis, Brit. Ent., pl. 96: designates Atalanta as type.

---

* Kirby (Syn. Cat. 612) strangely gives Hübner's malvæ (Eur. Schmett. 450-1) as a synonyme of sidæ, and not of alceæ.

1837. Sodoffsk., Bull. Mosc. x. 80: proposes to change the spelling of the name to Phanessa.
1840. Westw., Gen. Syn. 87: specifies Io as type.
1848. Doubl., Gen. Diurn. Lep. 98: Io, urticæ, and others.
1861. Feld., Neues Lep. 12: divides the group into five sections, placing urticæ in the third and Io in the fifth.
1871. Kirb., Syn. Cat. 181: employs it in Doubleday's sense, but subsequently (p. 648) treats it as a synonyme of Nymphalis.
1872. Scudd., Syst. Rev. 21: specifies Atalanta as type.
1872. Crotch, Cist. Ent. i. 66: would drop the name as synonymous with Nymphalis.

See Ammiralis, Bassaris, Pyrameis, and Cynthia.

1079. VICTORINA.

1840. Blanch., Hist. Ins. iii. 447: **Steneles.** Sole species, and therefore type.

Subsequently used in same sense by Doubleday, Westwood, Felder, and Kirby.

1080. VILA.

1871. Kirb., Syn. Cat. 217: **Azeca,** Mariana, Emilia, Stalachtoides.
1873. Ib., Zoöl. Rec. for 1871, 360: specifies Azeca as type, that having been the type of Olina, which this name is intended to supplant.

1081. XANTHIDIA.

1829-30. Boisd.-LeC., Lép. Am. Sept. 48: Delia, jucunda, Lisa, Nicippe.
1832. Boisd., Voy. Astrol. 59: Smilax, puella.
1833. Ib., Ann. Mus. Hist. Nat. ii. 168; Brigitta (pulchella) and others.

Delia, with which jucunda and Lisa are strictly congeneric, was taken in 1870 as type of Eurema (1816), so that Nicippe must be taken as the type of Xanthidia. See also Abœis.

1082. XANTHOCLEIS.

1870. Boisd., Lép. Guat. 30: Psidii, Themisto, **Ædesia** (Ædessa), and a MS. species.

Psidii and Themisto are congeneric but distinct from Ædesia; and Psidii is already the type of Thyridia; so that Ædesia must be taken as the type. See Aprotopos.

1083. Xanthotænia.

1857. Westw., Trans. Ent. Soc. Lond. [N. S.] iv. 187: **Busiris**.
Sole species, and therefore type.
1871. Kirb., Syn. Cat. 238: the same.

1084. Xenandra.

1865. Feld., Reise Novara, 304: **Heliodes**. Sole species, and therefore type.
1867. Bates, Journ. Linn. Soc. Lond. ix. 427: Helius, Heliodes (Helioides).
1871. Kirb., Syn. Cat. 301: the same.

1085. Xenica

1851. Westw., Gen. Diurn. Lep. ii. 387: **Achanta**, Klugii (Singa), Abeona, Lathoniella.
1858. Horsf.-Moore, Cat. Lep. E. Ind. Co. i. 228: employ it for Achanta only, which thereby becomes the type.
1867. Butl., Ent. Monthl. Mag. iii. 279: Abeona and Joanna.
1868. Ib., Cat. Sat. 70: specifies Abeona as type. See Heteronympha.
1871. Kirb., Syn. Cat. 76: uses it for some of Westwood's species and others, not including either of Butler's.
See Geitoneura.

1086. Xois.

1865. Hewits., Trans. Ent. Soc. Lond. [3] ii. 282: **Sesara**. Sole species, and therefore type, as stated by Butler and used by Kirby.

1087. Ypthima.

1816. Hübn., Verz. 63: Cassus (Casse), Hippia, Manto, Tyndarus (Cleo), **Philomela**.
1844. Doubl., List Br. Mus. 138: employs it for Philomela and others.
1851. Westw., Gen. Diurn. Lep. 394: makes a similar use of it, so that Philomela becomes the type.
1868. Butl., Ent. Monthl. Mag. iv. 196: designates Lisandra (Philomela) as type.
1871. Kirb., Syn. Cat. 94: makes a similar use of it.

1088. Zaretis.

1816. Hübn., Verz. 49: **Isidora**, Bisaltide (Polybete).
Isidora may be taken as the type.

1089. ZEGRIS.

1836. Ramb., Ann. Soc. Ent. Fr. v. 581 : **Eupheme.** Sole species, and therefore type.
1836. Boisd., Spec. gén. 552 : employs it for the same and others.
1847. Doubl., Gen. Diurn. Lep. 52 : the same.
1870. Butl., Cist. Ent. i. 39, 54 : specifies Eupheme as the type.

1090. ZELIMA.

1807. Fabr., Ill. Mag. vi. 279 : **Pylades.** Sole species, and therefore type.
1820. Billb., Enum. Ins. 81 : proposes, without reason, to supplant this name by Ailus (q. v.).

1091. ZELOTÆA.

1867. Bates, Journ. Linn. Soc. Lond. ix. 381 : **Phasma,** dubia, Achroa.
1871. Kirb., Syn. Cat. 310 : the same.

Phasma may be taken as the type.

1092. ZEMEROS.

1836. Boisd., Spec. gén., pl. 5 C.: **Flegyas** (Allica). Sole species, and therefore type.

Used in same sense by subsequent authors.

1093. ZEONIA.

1832-33. Swains., Zoöl. Ill. ii. 111 : **Faunus** (Heliconides). Sole species, and therefore type.

Used in same sense by Boisduval, Doubleday, Westwood, Bates, and Kirby. See Chorinea and Rodinia.

1094. ZEPHYRUS.

1816. Dalm., Vetensk. Acad. Handl. xxxvii. 62, 90 : all the species quoted under Aurotis, Heodes, and Cyaniris (q. v.), these being the three sections into which he divides this group. Betulæ is specified as the type.
1820. Dalm. in Billb., Enum. Ins. 80 [Zephyrius] : employs it for betulæ and others.
1832. Gray, Griff. An. Kingd., pl. 58 [Zephyrius] : uses it for Amor.
1842-44. Guér., Iconogr. Règne An. 490, pl. 81 [Zephyrius] : the same.

1853. Wallengr., Rhop. Scand. 178: employs it for quercus and betulæ.
1871. Kirb., Syn. Cat. 402: uses it for the same and others.
See Aurotis.

1095. ZERENE.

1816. Hübn., Verz. 97: Croceus (Hyale), Erate, Hyale (Palæno), Phicomene, **Cesonia.**
1850. Steph., Cat. Brit. Lep. 3 [Xerene]: employs it for Hyale alone; but this cannot be taken as the type, since it had previously been made the type of Eurymus. See also Colotis.
1862. Scudd., Proc. Bost. Soc. Nat. Hist. ix. 103: employs it for Cesonia (Cœsonia) and Eurydice, wherefore Cesonia is type.
1872. Ib., Syst. Rev. 38: specifies Cesonia (Cæsonia) as the type.
1872. Grote, Can. Ent. iv. 215: says that this group, being synonymous with Colias, cannot be used, and that Megonostoma (q. v.) should be employed; but it is not strictly synonymous with what Grote means by Colias.

1096. ZERITIS.

1836. Boisd., Spec. gén., pl. 6 C.: **Neriene.** Sole species, and therefore type.
1847. Doubl., List Br. Mus. 56: employs it for the allied species Thero, and for others, but not for Neriene.
1849. Luc., Expl. Alg. Zoöl. iii., pl. 1: Siphax, a wholly different insect. See Cigaritis.
1852. Westw., Gen. Diurn. Lep. 500: uses it for fourteen species, among them Neriene and Thero.
1857. Wallengr., Rhop. Caffr. 46 [Zerythis]: uses it for Protumnus (Basuta).
The name is very close to Zarctis (Hübn., Lep. 1816).

1097. ZERYNTHIA.

1816. Ochs., Schmett. Eur. iv. 29: **Polyxena,** Rumina (Medesicaste, Rumina).
1822–26. Hübn., Exot. Schmett. ii.: uses it for Ogina, an entirely different insect.
1835. Herr.-Schaeff., Nomencl. Ent. i. 4: employs it in Ochsenheimer's sense.

1837. Sodoffsk., Bull. Mosc. x. 82: suggests that it should be spelled Zerinthia.
    Polyxena may be taken as type. See also Eugraphis and Thais.

1098. Zesius.

1816. Hübn., Verz. 77: Phœomallus, **Chrysomallus**.
    Chrysomallus may be taken as the type.

1099. Zethera.

1861. Boisd. in Feld., Neues Lep. 26: **Pimplea**. Sole species, and therefore type, as stated by Butler.
1871. Kirb., Syn. Cat. 45: employs it in the same sense.
    See Amechania.

1100. Zetides.

1816. Hübn., Verz. 85: **Sarpedon**, Eurypylus, Ægistus.
    Sarpedon may be taken as the type. See Chlorisses.

1101. Zeuxidia.

1822–26. Hübn., Exot. Schmett. ii.: **Luxerii**. Sole species, and therefore type.
1844. Doubl., List Br. Mus. 114: the same.
1851. Westw., Gen. Diurn. Lep. 327: the same and others. Westwood gives Aglaura Boisd. MS. as a generic synonyme.
1871. Kirb., Syn. Cat. 115: uses it in the same sense.

1102. Zipætis.

1863. Hewits., Exot. Butt. iii. 100: **Saitis**, Scylax.
1865. Herr.-Schaeff., Prodr. i. 63: the same.
1868. Butl., Ent. Monthl. Mag. iv. 194; Cat. Sat. 98: specifies Saitis as type.

1103. Zonaga.

1820. Billb., Enum. Ins. 78: **Biblis**. Sole species, and therefore type.
    See Didonis and Biblis.

1104. Zophoessa.

1849. Doubl., Gen. Diurn. Lep. pl. 61: **Sura**. Sole species, and therefore type.
1851. Westw., Gen. Diurn. Lep. 362: the same.
1868. Butl., Ent. Monthl. Mag. iv. 195; Cat. Sat. 108: specifies Sura as type.
1871. Kirb., Syn. Cat. 40: employs it in the same sense.

The following species of butterflies, mentioned as types of genera, were unpublished at the time of the issue of Kirby's Catalogue:—

Thaldina,	of Armandia (Blanch.), 1871.		Poweshiek,	of Oarisma (Scudd.), 1872.
tractipennis,	Arteurotia (Butl.-Druce), 1872.		[oolitica,	Palæontina (Butl.), 1873.]
Lidderdalii,	Bhutanitis (Atkins.), 1873.		Leda,	Periplysia (Gerst.), 1871.
Juventus,	Callimormus (Scudd.), 1872.		Aetta,	Pteronymia (Butl.-Druce), 1872.
Leonata,	Drucina (Butl.), 1872.		Reynesii,	Satyrites (Scudd.), 1872.
Darwinia,	Minacræa (Butl.), 1872.		Hermina,	Scalidoneura (Butl.), 1871.

ADDENDA. — (MARCH, 1875.)

46. ALCIDIS. — This name was introduced by an accidental error. Liris is not a butterfly, and was not given as one by Felder.

152. AUROTIS. — Add: 1835. Vill.-Guén., Lép. Eur. 36: employs it for roboris (Evippus). — 1862. Kirb., Man. Eur. Butt. 87: roboris.

256 bis. CHORTOBIUS.*

1859. [Guén. in] Doubl., List Brit. Lep. Ed. 2: Typhon (Davus), Pamphilus. Fide Kirby in litt. Falls before Coenonympha.

302. CUPIDO. — Add: 1870. Kirb., Journ. Linn. Soc. Zoöl. x. 499: says, "The true type appears to be Alsus;" because, he writes me in explanation, "Schrank confounds Alsus and Argiades as sexes under his Puer," the name Puer being presumed to have suggested Cupido; but this seems to me rather strained.

305. CYANIRIS. — Add: 1835. Vill.-Guén., Lép. Eur. 19: employ it for Corydon, Argiolus, and others.

492. HÆMONIDES. — Mr. Kirby writes me: "Cramer figures two species as Cronis, one a Castnian, the other a Pierid. Boisduval and I take this to be a case of mimicry; but Butler considers both figures to represent the Castnian."

510. HELIOCHROMA. — 1870. Butl., Lep. Exot. 70: says, "The genus Heliochroma will, I think, have to sink into a section of Hesperocharis. I can find no constant structural characters by which to separate it."

581. ITHOMIA. — With regard to the text of Hübner's Sammlung exotischer Schmetterlinge, it may be remarked that the twelve species described in it are all figured in the first volume, and all referred to in the Index of 244 plates. And inasmuch as in every case of alteration of the specific name, the Index is followed, we may conclude the text of the Sammlung to be posterior to, or most probably nearly synchronous with, the Index, namely, 1822. The genus in which Dianasa is placed is spelled Eicides, as in the Index, and not Eueides as in the Verzeichniss; and further proof that it is later than the Verzeichniss is found in the entire absence of one of the species (and its generic name) from the latter, — Heliochlaena Leucosia.

633. LIMENITIS. — Mr. Kirby writes me that the Camilla of early British authors is not that of Fabricius, and cannot therefore be taken as type. But inasmuch as it was a strictly congeneric insect (Sibylla), the question is not affected by this fact.

755 bis. NYMPHA.*

1838–9. Krause, Faun. Thur., wrapper parts 4, 5: proposes it to include all the European Nymphales. Mr. Kirby, from whom this information is derived, appears sometimes to write it Nympha, sometimes Nymphæ. The latter form would be inadmissible in a generic name, and is also given earlier by Borkhausen (Eur. Schmett., Einl. xvii.) as a name for the whole family. Mr. Kirby adds: "On p. 85, populi is clearly, as I think, indicated as type." In that case the name would fall before Najas.

861. PURISSURA. — Add: 1871. Butl., Trans. Ent. Soc. Lond. 171: says the insect upon which he intended to found this genus was Ægis (Illana), which at the time he wrongly identified as Cynis.

www.ingramcontent.com/pod-product-compliance
Lightning Source LLC
Chambersburg PA
CBHW020909230426
43666CB00008B/1370